Cahiers de Logique et d'Épistémologie
Volume 22

Soyons Logiques

Let's be Logical

Volume 17
Argumentation et engagement ontologique. Être, c'est être choisi
Matthieu Fontaine

Volume 18
L'arbre du *Tractatus*
Luciano Bazzocchi. Traduit de l'Italien par Jean-Michael Luccioni

Volume 19
L'émergence de la Presse Mathématique en Europe au 19ème Siècle.Formes éditoriales et études de cas (France, Espagne, Italie, et Portugal)
Christian Gerini and Norbert Verdier, eds.

Volume 20
Entre l'orature et l'écriture. Relations croisées
Charles Zacharie Bowao and Shahid Rahman, eds.
Préface de Cristian Bernar and Marcel Nguimbi

Volume 21
La sémantique dialogique. Notions fondamentales et éléments de metathéorie
Nicolas Clerbout

Volume 22
Soyons Logiques / Let's be Logical
Amirouche Moktefi, Alessio Moretti et Fabien Schang, directeurs de publication.

Cahiers de Logique et d'Épistémologie Series Editors
Dov Gabbay   dov.gabbay@kcl.ac.uk
Shahid Rahman   shahid.rahman@univ-lille3.fr

Assistance Technique
Juan Redmond   juanredmond@yahoo.fr

Comité Scientifique: Daniel Andler (Paris – ENS); Diderik Baetens (Gent); Jean Paul van Bendegem (Vrije Universiteit Brussel); Johan van Benthem (Amsterdam/Stanford); Walter Carnielli (Campinas-Brésil); Pierre Cassou-Nogues (Lille 3 – UMR 8163-CNRS); Jacques Dubucs (Paris 1); Jean Gayon (Paris 1); François De Gandt (Lille 3 – UMR 8163-CNRS); Paul Gochet (Liège); Gerhard Heinzmann (Nancy 2); Andreas Herzig (Université de Toulouse – IRIT: UMR 5505-NRS); Bernard Joly (Lille 3 – UMR 8163-CNRS); Claudio Majolino (Lille 3 – UMR 8163-CNRS); David Makinson (London School of Economics); Tero Tulenheimo (Helsinki); Hassan Tahiri (Lille 3 – UMR 8163-CNRS).

Soyons Logiques

Let's be Logical

Amirouche Moktefi,
Alessio Moretti
et
Fabien Schang
directeurs de publication

© Individual authors and College Publications 2016
All rights reserved.

ISBN 978-1-84890-090-5

College Publications
Scientific Director: Dov Gabbay
Managing Director: Jane Spurr

http://www.collegepublications.co.uk

Printed by Lightning Source, Milton Keynes, UK

All rights reserved. No part of this publication may be reproduced, stored in a retrieval system or transmitted in any form, or by any means, electronic, mechanical, photocopying, recording or otherwise without prior permission, in writing, from the publisher.

Table of Contents

Préface : Soyons Logiques !... vii
Amirouche Moktefi, Alessio Moretti et Fabien Schang

Let us be Antilogical: Anti-Classical Logic as a Logic.......................... 1
Jean-Yves Béziau and Arthur Buschbaum

Les Oppositions Modales dans la Logique d'Al Fārābi...................... 11
Saloua Chatti

Formalism. The Success(es) of a Failure... 31
Liesbeth De Mol

La Quadrature du Carré 49
George Englebretsen

Chose, Concept et Molécularisme : autour de la proposition
des *Principles* de Russell 65
Philippe Gac

Bloody-Minded Revolutions (An Essay Inspired by the
Wittgensteinian *Jaha!*)... 83
Katarzyna Gan-Krzywoszyńska and Piotr Leśniewski

Formal Logic and Ancient Mathematical Reasoning 97
Pierluigi Graziani and Maurizio Colucci

Coherence and Contradiction... 117
John T. Kearns

Approaches Phénoménologiques de la Vérité
Mathématique.. 129
Frédéric Patras

Modal Truths from an Analytic-Synthetic Kantian
Distinction 149
Francesca Poggiolesi

How to Hintikka a Frege 161
Fabien Schang

The Square of Opposition that Was Lost..173
Andrew Schumann

What Is a Refutation System? 189
Tomasz Skura

Préface : Soyons Logiques !
Amirouche Moktefi, Alessio Moretti et Fabien Schang

'Soyons logiques!'. Une exhortation à double sens, puisque la logique peut s'entendre aussi bien comme une disposition acquise de l'esprit qu'une technique complexe à acquérir. Par ailleurs, l'ouvrage comporte deux approches différentes de la discipline logique : une approche rétrospective destinée à rappeler quelques étapes essentielles dans l'histoire de la logique, et une approche prospective où seront exposées plusieurs méthodes de logique contemporaine. En cela, l'ouvrage propose un voyage intellectuel à travers plusieurs époques et plusieurs objectifs. Car si la logique a bien un trait qui la rend si singulière, c'est son caractère universalisant et sa réputation de 'science des sciences' qui fait d'elle un exercice à la fois propédeutique et normatif. Dans un cas comme dans l'autre, il va de soi que l'ouvrage ne prétend pas à l'exhaustivité. Néanmoins, il est espéré que ce volume contribuera à la réflexion sur ce que la logique fut, est ou sera peut-être dans quelques années à la lumière des recherches actuelles.

Plutôt que d'opposer une partie historique à une partie philosophique, le lecteur est convié à une double mise en perspective de la logique. Une première dimension abordée par les articles de cette collection est d'ordre normatif et renvoie à l'idée d'une théorie universelle de la logique. Question complexe, mais incontournable : y a-t-il une seule et même logique à la base de tous les raisonnements humains ? Si tel est le cas, de quels raisonnements s'agit-il, et dans quelle mesure peut-on garantir leur validité ? La relation instaurée entre la logique et les mathématiques va semble-t-il dans ce sens, quelle que soit l'issue de la question portant sur le rôle de fondement attribué à la logique dans le domaine des mathématiques. La seconde dimension est d'ordre descriptif, où la logique permet une analyse rigoureuse d'idées ou concepts empruntés au domaine de la philosophie. La sous-discipline qui en résulte : la 'logique philosophique', sera illustrée à travers un ensemble d'articles portant sur les logiques dites 'modales'. Dans celles-ci, la logique moderne classique est enrichie dans son vocabulaire et passe d'un discours unique sur la vérité à une série de modes de jugements complémentaires (selon que la vérité est sensible à des contextes temporels, épistémiques, déontiques, etc.). La rigueur est toujours de mise dans cette approche, bien entendu ; en revanche, l'universalité du système logique proposé n'y est pas affirmée puisque le logicien n'a d'autre prétention ici que d'organiser un ensemble de raisonnements plus limités dans leur portée.

Un philosophe a dit : 'N'écoute pas ce que je dis, regarde ce que je fais'. Gageons que le lecteur n'en pensera pas moins. Mieux que des paroles générales et de vagues promesses à son adresse, une brève présentation des articles à venir, organisés par ordre alphabétique des auteurs, devrait lui donner ainsi une idée plus précise du parcours qui l'attend.

Dans « Let us be antilogical : Anti-classical logic as a logic », Jean-Yves Béziau et Arthur Buchsbaum s'interrogent sur les frontières de la discipline logique à travers un cas d'étude aussi particulier que singulier : la logique 'anti-classique', c'est-à-dire un système logique qui est le complémentaire de la logique classique. Après avoir inspecté ce système, les auteurs en concluent que la perte de deux de ses propriétés métathéoriques fondamentales n'empêche pas la logique anti-classique de demeurer une logique à part entière. C'est de cette généralité que l'article traitera, pour conclure la réflexion d'ensemble.

Dans « Les oppositions modales dans la logique d'Al Farabi », Saloua Chatti nous ramène aux sources d'une tradition logique mal connue : la logique dite 'arabe' (ou islamique), inspirée par les travaux d'Aristote mais porteuse également d'un corpus à part entière. Dans cet article, Chatti propose une lecture comparative des modalités de nécessité et de possibilité telles qu'elles furent conçues par Aristote et Al Farabi. Non seulement l'originalité de ce dernier est avérée, mais l'auteur étend son analyse à l'époque médiévale et trouve plusieurs affinités entre l'analyse des modalités d'Al Farabi et celle de Buridan.

Un autre clin d'œil historique est proposé dans le troisième article, « Formalism : The success(es) of a failure », où Lisbeth De Mol rappelle la contribution du logicien polonais Emil Post à une meilleure compréhension du concept de 'forme'. Essentiels pour la logique, mais restreints dans leur champ d'application par les résultats ultérieurs de Gödel, les travaux de Post apportent, selon l'auteur, aujourd'hui encore un éclairage pertinent sur le rôle du formalisme, notamment à travers l'exercice de computation dans le domaine de l'informatique.

L'attention est portée ensuite sur la logique traditionnelle. Dans « La quadrature du carré », George Englebretsen réinvestit la question du carré des oppositions issu de la logique d'Aristote. Loin de déconsidérer cet outil d'analyse logique, Englebretsen y voit au contraire un moyen de mieux comprendre les ressources du langage naturel. Mieux, il propose une réhabilitation de la logique des termes, afin de donner une lecture plus claire des concepts d'opposition et de répondre au préjugé de la logique moderne en vertu duquel la théorie des oppositions serait fatalement invalidée par plusieurs cas de raisonnements consacrés par la logique des prédicats. Plutôt que de céder à une telle posture, l'auteur nous invite à penser autrement la logique des oppositions, par le biais d'une algèbre simple et plus conforme à la logique des termes.

Dans « Chose, concept et molécularisme autour de la proposition des Principles de Russell », Philippe Gac revient lui aussi aux sources d'une notion fondamentale de la logique moderne : la 'proposition', à travers une relecture problématique du concept tel qu'il a été façonné par Bertrand Russell. Pour mesurer l'impact que l'analyse de ce dernier a pu avoir sur la pratique moderne de la logique, Gac explique que la proposition selon Russell avait une portée ontologique dont l'expression visait à décrire la nature des choses elles-mêmes. Au cours d'une réflexion lucide, l'auteur ne manque pas de rappeler l'engagement atomiste de Russell et le combat acharné que le Britannique mena contre l'idéalisme de son époque. Pour conclure, Gac demande si une autre analyse logique du langage est possible au sein de la philosophie de Russell. Le thème du molécularisme est abordé à ce titre.

Dans « Bloody-minded revolutions (an essay inspired by the Wittgensteinian *Jaha!*) », Katarzyna Gan et Piotr Lesniewski nous proposent une lecture (r)évolutionniste de la discipline logique dans son ensemble. A la question de savoir si les théories logiques croissent et meurent selon un cycle somme toute naturel, les auteurs tentent de répondre par l'affirmative en comparant deux tournants décisifs dans l'histoire de la logique : le 'tournant mathématique' du début du vingtième siècle, puis le 'tournant pratique' opéré un siècle plus tard. Pour donner de la hauteur au problème posé, Gan et Lesniewski rappellent l'enseignement professé par Franz Brentano : seul un retour à la philosophie permet un aperçu général de la logique et de son évolution. Il s'agira de savoir si, à l'aune des propos de Brentano, le tournant pratique observé en logique constitue bien le signe de son déclin.

Pour revenir sur la relation entre logique et mathématiques, la notion de forme est de nouveau invoquée dans « Formal logic and ancient mathematical thinking ». Dans ce septième article, Pierluigi Graziani et Maurizio Colucci s'intéressent aux mérites et limites du formalisme moderne appliqué au domaine des mathématiques anciennes. Pour juger de la pertinence d'un outil moderne à travers l'histoire de la pensée, les auteurs ont porté une attention particulière au cas des postulats d'Euclide et opposent à la méthode formelle d'autres approches telles que le raisonnement visuel et la démonstration par instanciation. La valeur explicative du formalisme sera justifiée au final, à titre de méthode d'analyse unificatrice pour l'ensemble des mathématiques anciennes.

Dans « Coherence and contradiction », John T. Kearns met en évidence un autre aspect de la pratique logique : sa dimension dite 'illocutoire', en vertu de laquelle la signification d'un acte de discours exige plus qu'une simple analyse standard des conditions de vérité du contenu propositionnel. Dans cette perspective, Kearns développe un système de logique illocutoire composé à la fois de constantes logiques et de règles de rationalité. Tout d'abord, les opérateurs classiques de l'affirmation et de la négation sont remplacés ici par des actes d'assertion et de dénégation. Puis le principe de non-contradiction est revu à la lumière de la théorie des actes de discours sous-jacente et réhabilité à ce titre dans les termes d'un critère de 'cohérence'. L'auteur précise que des croyances incohérentes peuvent exister, en raison d'un manque d'informations logiques suffisantes chez le locuteur ; mais cela n'implique en rien le rejet du principe de non-contradiction, dès lors que les règles illocutoires du discours sont ajoutées à l'appareil locutoire classique.

L'interaction entre logique et philosophie se prolonge dans l'article de Frédéric Patras qui propose une lecture hardie de la notion de vérité. Dans « Approches phénoménologiques de la vérité mathématique », Patras propose une relecture phénoménologique de quelques notions centrales en philosophie de la logique : la vérité et les théories explicatives qui la sous-tendent, mais aussi le statut de l'énoncé et des preuves. L'ensemble conduit, selon l'auteur, à porter un autre regard sur la logique mathématique et l'informatique théorique.

Dans « Modal truths from an analytic-synthetic Kantian distinction », Francesca Poggiolesi réinvestit une autre question illustre posée par la philosophie de la logique : la nature des vérités logiques. A rebours de la tradition héritée de Kant, pour qui les vérités logiques étaient de nature analytique, Poggiolesi reprend à son compte les travaux de Hintikka sur la question et propose non seulement une double définition de l'analyticité, mais aussi une extension du point de vue de Hintikka au domaine des vérités logiques modales. Les résultats récents obtenus en théorie de la preuve constituent, selon l'auteur, un moyen efficace pour réfuter le postulat kantien en la matière.

A propos de Hintikka, l'article suivant s'intéresse à la principale contribution du logicien finlandais : la logique épistémique, en particulier la version 'statique' du système basée sur l'analyse formelle des concepts de connaissance et de croyance. Dans « How to Hintikkize a Frege », Fabien Schang propose de porter un regard différent sur cette logique philosophique et de la considérer du point de vue inverse de la philosophie de la logique. Dans un premier temps, deux théories de la signification sont décrites et associées à deux théories rivales de la compétence linguistique. Dans un second temps, Schang en tire la conclusion que la logique épistémique de Hintikka constitue une sorte d'internalisation du sens, par l'introduction d'opérateurs modaux épistémiques au sein d'un langage-objet. A cet égard, considérer la signification comme la résultante d'une compétence linguistique fait de la logique épistémique rien de moins qu'une logique du sens et de la compréhension unifiés.

Dans « The square of opposition that was lost », Andrew Schuman revient sur le thème de la théorie des oppositions et déclare que celle-ci n'est pas close dans la mesure où un autre ensemble de dualités existe au-delà du formalisme hérité du corpus logique d'Aristote. Sur la base de ces résultats, l'auteur propose à la fois une syllogistique non-standard des propositions synthétiques et une explication des verbes performatifs inspirée de la théorie des actes de discours illocutoires.

Pour conclure, la notion d'opposition est encore mise à l'honneur par le biais d'une redéfinition générale de la discipline logique. Dans « What is a refutation system ? », Tomasz Skura remplace le concept central de déduction par celui de réfutation, substituant ainsi à l'approche tarskienne de la relation de conséquence une relecture des méthodes d'analyse logique en termes de rejet. De nouvelles méthodes sont élaborées par la suite, sur la base des modèles et tableaux sémantiques. L'utilité de ce renversement de perspective est justifiée par l'auteur, en même temps qu'il en expose les aspects techniques.

Les auteurs des contributions rassemblées dans ce volume ont été invités à présenter leurs textes de sorte qu'ils soient accessibles à un large lectorat de logique. Ces textes ont, par ailleurs, été soumis à une évaluation par rapporteurs anonymes. Nous souhaitons donc remercier les auteurs et les rapporteurs pour leur contribution à la réalisation de cet ouvrage et pour leur patience tout au long sa longue préparation. Nous remercions également Jane Spurr pour son aide à la finalisation du volume. A noter enfin que la publication de cet ouvrage a bénéficié du soutien financier de l'Université de Strasbourg.

Amirouche MOKTEFI
Ragner Nurkse School of Innovation and Governance
Tallinn University of Technology, Estonia

Alessio MORETTI
Chercheur indépendant

Fabien SCHANG
National Research University Higher School of Economics

Let us be Antilogical:
Anti-Classical Logic as a Logic

JEAN-YVES BÉZIAU AND ARTHUR BUCHSBAUM

1 Challenging Logicality

To be logical seems important, but what does it mean exactly? When are we logical and when are we illogical? Is it logical to think that $2 + 2 \neq 5$, that God does not exist, that it is impossible to go at a speed higher than light? It can be logical or not, it depends on what kind of thinking is behind, for which *reasons* we think so. We can also think logically or illogically the contrary, i.e. that $2 + 2 = 5$, that God exists, that it is possible to go at a speed higher than light.

It is important to make the distinction between logic as reasoning and logic as the theory of reasoning, a distinction that can be expressed as the difference *logic versus Logic* (for more details about this distinction, see [11]). René Descartes and Blaise Pascal were against a theory of logic such as syllogistic, but they were not against Logic, i.e. against being logical. Pascal wrote: "It is not Barbara and Baralipton that constitute reasoning. The mind must not be forced; artificial and constrained manners fill it with foolish presumption, through unnatural elevation and vain and ridiculous inflation, instead of solid and vigorous nutriment." [18]

For them being logical was to follow such basic principles such as:

- Never to accept anything for true which I did not clearly know to be such; that is to say, carefully to avoid precipitancy and prejudice, and to comprise nothing more in my judgment than what was presented to my mind so clearly and distinctly as to exclude all ground of doubt. (Descartes, [17])

- Not to employ in the definition of terms any words but such as are perfectly known or already explained. (Pascal, [18])

For them being logical is not to follow the rules of an artificial system such as syllogistic. But can we say the same nowadays after the Boolean revolution, the mathematization of logic, that has not only provided a more accurate description of reasoning but also transformed our way of reasoning, pushing the limits of rationality and developing artificial intelligence? There are thousands of systems of logic, from linear logic to erotetic logic, from alethic modal logic to turbo polar logic. It seems that it is possible to develop a logic for anything. Everything has its logic, everything is logical... There are even logics for reasoning with contradictions, the so-called "paraconsistent logics" (see e.g. [12]), which is absurd in the light of the principle of non-contradiction, which has been considered as the basis of logicality during many centuries.

To understand the nature of the logical in modern times, one has to inquire about the general nature of logical systems. This is what we will do here using anti-classical logic as a guide.

2 From Tarskian logical structures to anti-classical logic

Following the spirit of modern mathematics, we can say that a logic is a structure. But what kind of structure? The notion which has emerged is due to Alfred Tarski (his first ideas appear in [23], were reprinted in [13] with comments in [28]; see also [8]). It is a set with a consequence relation on it:

$$\mathcal{L} = \langle \mathbb{L}, \vdash_{\mathcal{L}} \rangle.$$

The consequence relation $\vdash_{\mathcal{L}}$ is a relation between a set of premises T, elements of $\mathbb{L}$, leading to a conclusion a, also an element of $\mathbb{L}$. Classical propositional logic, first-order classical logic and second-order classical logic can be seen as such logical structures. The many systems of intuitionistic, many-valued, paraconsistent and modal logics can also be seen in this way. However Tarski chose some axioms that exclude from this realm several kinds of logic.

The three Tarskian axioms are the following:

- **T1**: If $a \in T$, $T \vdash_{\mathcal{L}} a$.

- **T2**: If $T \vdash_{\mathcal{L}} a$ and $T \subseteq U$ then $U \vdash_{\mathcal{L}} a$.

- **T3**: If $T \vdash_{\mathcal{L}} a$ and $U, a \vdash_{\mathcal{L}} b$ then $T, U \vdash_{\mathcal{L}} b$.

These axioms say nothing explicitly about the logical operators: connectives, modalities, quantifiers, etc. So they are all welcome unless they indirectly contradict these axioms. These three axioms can be seen as respectively expressing reflexivity (**T1**), monotonicity (**T2**) and transitivity

(**T3**), some of the so-called "structural" properties of a logic. The most famous excluded logics from this framework are the non-monotonic logics promoted by John McCarthy and people working in artificial intelligence. Their argument for supporting such extravaganza is based on penguins and other empirical phenomena. Is it possible to find other creatures or objects that could serve as a basis for the rejection of reflexivity and/or transitivity? Yes, there are plenty of them. But rejection of Tarski's axioms can be based not only on empirical data but also on theoretical reasons.

There is a natural structure which obeys none of the Tarskian axioms, this is *anti-classical logic*. If we consider classical logic as the structure

$$\mathcal{K} = \langle \mathbb{K}, \vdash_{\mathcal{K}} \rangle,$$

anti-classical logic is then defined as

$$\overline{\mathcal{K}} = \langle \mathbb{K}; \vdash_{\overline{\mathcal{K}}} \rangle,$$

with the following definition of the anti-classical consequence relation:

$$T \vdash_{\overline{\mathcal{K}}} a \text{ iff } T \nvdash_{\mathcal{K}} a.$$

Here are examples showing that anti-classical logic does not obey Tarskian axioms:

- **T1**: $e \nvdash_{\overline{\mathcal{K}}} e$ (e being any proposition).
- **T2**: $d \vdash_{\overline{\mathcal{K}}} e$ and $\{d, e\} \nvdash_{\overline{\mathcal{K}}} e$ (d, e atomic).
- **T3**: $d \vdash_{\overline{\mathcal{K}}} e$ and $e \vdash_{\overline{\mathcal{K}}} d$ but $d \nvdash_{\overline{\mathcal{K}}} d$ (d, e atomic).

There are two issues:

- **I1**: Can we consider anti-classical logic as a logical structure despite the fact that it does not obey any Tarskian axioms?

- **I2**: Can we say that anti-classical logic describes a logical way of reasoning?

These two issues are intertwined. If we consider anti-classical logic as a structure of type $\mathcal{L} = \langle \mathbb{L}, \vdash_{\mathcal{L}} \rangle$, we can see $\vdash_{\mathcal{L}}$ as a road leading from some hypotheses T to a conclusion a, what should we specify for this road to be logical? Can be put no axiom on $\vdash_{\mathcal{L}}$? We can indeed argue in favor of axiomatic emptiness from some theoretical reasons.

Axiomatic emptiness has been promoted in the field of universal algebra by Garrett Birkhoff. He has developed a purely conceptual approach defining an algebra as a set with a family of operators obeying no axioms (see [15], [16]), by contrast with his predecessors Sylvester [22] and Whitehead [24], who were looking for some universal axioms for algebraic structures. For Birkhoff an abstract algebra is just a structure of type $\mathcal{A} = \langle \mathbb{A}, f_i \rangle$. This definition is enough to start working: in particular we can define the notion of subalgebra and morphism, they don't depend on any axioms. To have no axioms is no problem, it is in fact an advantage from a theoretical viewpoint, it allows to develop a smooth and universal theory. The same strategy can be applied to logical structures, this is the way to *universal logic* (cf. [2], [3], [7]).

From this perspective we can admit as logical structures any kind of structure of type $\mathcal{L} = \langle \mathbb{L}, \vdash_{\mathcal{L}} \rangle$. We have then two extreme cases, the logic in which nothing is a consequence of nothing, let us call it $\mathcal{L}_\emptyset$ (nickname: "zerologic") and the one in which everything is a consequence of everything, let us call it $\mathcal{L}_0$ (nickname: "cathologic"). If we consider a fixed domain $\mathbb{L}$, any logic on this domain is included in the cathologic on $\mathbb{L}$, includes the zerologic on $\mathbb{L}_\emptyset$, and it has an anti-logic which is also part of the network of logics defined on $\mathbb{L}$. We have a nicely structured class of logics. It is furthermore possible to use the square of opposition to describe the relations between logics and antilogics, as shown in [14] and [1].

If Eloise says that this is nonsense, Abelard will reply to her that this is not only nonsense, but general abstract nonsense. But we can be less Abelardian than Abelard arguing that anti-classical logic is a concrete logic. In fact it is possible to construct a semantics for anti-classical logic and also a proof-theoretical system, models and proofs being the two teats of modern logic, that sounds good and may produce a nice milky logic, on the basis of a completeness theorem establishing a link between these two sources of productions.

One of the godfathers of modern logic, the Polish logician Jan Łukasiewicz, has worked in this direction by developing refutation systems, i.e. proof-systems generating all formulas of a logic which are not logically valid (for recent works of the Polish School on this subject see [20], [21] [26]). If one considers that a logical argument is something than can be decomposed step by step, where each step can be justified by the application of a rule, then anti-classical logic proofs can be considered as logical arguments.

3 Substitution and replacement

There are two important metalogical features which are not valid in anti-classical logic. These are the so-called *substitution theorem* and *replacement theorem*. In modern logic there has been some ambiguity about the status of these two features, which is manifest in the expressions used to call them. Sometimes one of these expressions is used for the other, despite the fact that these are two quite different features — only apparently similar, false cognates — and moreover these properties are not necessarily theorems.

3.1 You shall not substitute

Substitutivity says that if we replace all the occurrences of an atomic formula by an arbitrary formula in a valid reasoning, then we still have a valid reasoning. In the Polish school Łoś and Suszko, pursuing the work of Tarski, have considered substitutivity as a fourth axiom they have added to the three Tarskian axioms, this led to the so-called *structural consequence relations* (cf. [19], reprinted in [13], with comments in [27]).

This property has a fundamental philosophical significance, which can be traced back to Aristotle, who was already using schematic letter to express it, and which makes sense within his hylemorphic views, i.e. the distinction between form and matter. In logic this means that logicality is formal in the sense that logical truth does not depend on the meaning but only on the form of the reasoning. This leads to a conception of logic based on logical forms.

However this property can be rejected for various reasons, in particular if one wants to take seriously in account meaning within the logical kingdom. People emphasizing meaning sometimes prefer to go outside of the logical realm, this was the position of Wittgenstein in his second period, who was rejecting logical systems as meaningless. Some other meaningful people apparently want to stay within the logical realm, saying they are doing "informal logic", but they are not dealing with logical systems. However it is in fact possible to deal with logical systems taking into account meaning, and a way to do it is to consider non-structural logics in which the substitution property does not hold.

From the perspective of axiomatic emptiness of universal logic [10], a logic structure does not necessarily obey the substitution axiom of Łoś and Suszko, a logic can be non-structural but still be a structure! And in this case we can still speak of *formal logic* if we consider that the form is located at a higher level of abstraction (cf. [9]).

One has to be careful: a logic in which the substitution property does not hold is not necessarily a logic of meaning. A good example is anti-classical logic which is a logical structure which is non-structural but which is not more meaningful than classical logic, the best known meaningless logic.

Let's see counterexamples of substitutivity in anti-classical logic. If we consider two atomic propositions such as $e = $ *Snow is white* and $d = $ *God is blue*, in anti-classical logic it is possible to deduce that *Snow is white* from *God is blue*, on the other hand it is not possible to deduce that *Snow is white* from the proposition f according to which *The sun is red and the sun is not red*. This is due to the fact that this reasoning is valid in classical logic, so we have here a typical failure of the substitution property. Symbolically, we have that $d \models_{\overline{K}} e$, but $f \not\models_{\overline{K}} e$, since f is of the form $a \wedge \neg a$ and $a \wedge \neg a \models_{\overline{K}} e$.

Relevant logicians have also rejected the fact that from a contradiction anything follows, arguing that there are no meaningful connections in this case between the premises and the conclusion, and they have been paraconsistent for this reason. For them a way to try to catch meaning is to require that there are some atomic propositions in common between the premises and the conclusion. Anti-classical logic is not relevant, and it is also fully meaningless since, given two atomic propositions, one is a consequence of the other even if they have no common meaning.

In view of our above example, one may think that anti-classical logic is paraconsistent. This is a quite common deficient way of thinking, according to which a paraconsistent negation is any unary negation not obeying (one form) the ex falso sequitur quodlibet. But, as it has been stressed at length elsewhere (cf. [4] and [5]), a paraconsistent negation should also be defined positively, to be sure that we are talking about a kind of "negation", not an arbitrary unary connective. Thinking we are dealing with negation just because we are using the symbol "¬" is a symbol of the illusionism of symbolism. In the case of anti-classical logic, the use of the symbol "¬" is to keep in mind how the logic has been generated, but the connective denoted by "¬" in anti-classical logic is in no sense a negation.

This does not mean that there are no valid schemes of formula or consequence. For example, the proposition $e \wedge \neg e$ is valid in $\overline{K}$ and so is any substitution of it. In anti-classical logic, among the tautologies, we have those which are individual and those which are schematic. This is an interesting distinction that can be used for any non structural logic. Examples of non schematic tautologies in $\overline{K}$, besides atomic propositions, are formulas like $e \rightarrow (e \wedge d)$ or $(e \rightarrow d) \rightarrow (d \rightarrow e)$. In these two formulas if we substitute d for e then we have formulas which are not any more tautologies of $\overline{K}$. A schematic tautology of $\overline{K}$ is in fact an antilogy of K.

3.2 You shall not replace

The notion of logical equivalence can be defined for any consequence relation, independently of specific axioms. We say that two formulas a and b are logically equivalent in a logic $\mathcal{L}$ iff $a \models_{\mathcal{L}} b$ and $b \models_{\mathcal{L}} a$ This is abbreviated as $a \models\!\!\models_{\mathcal{L}} b$.

The replacement theorem says that if we replace an occurrence of a formula a by another formula equivalent to it b we preserve the consequence relation, in particular, if $a \dashv\vdash_{\mathcal{L}} b$, then $c \dashv\vdash_{\mathcal{L}} c(b/a)$, where c is a formula in which we have replaced a by b.

In anti-classical logic, the replacement theorem is not valid. We can consider the following counterexample: e and d are atomic formulas and they are logically equivalent since $e \dashv\vdash_{\overline{\mathcal{K}}} d$, but, if we replace the first occurrence of d by e in the formula $d \wedge \neg d$, then $d \wedge \neg d$ is not logically equivalent to $e \wedge \neg d$, because $d \wedge \neg d \not\equiv_{\overline{\mathcal{K}}} e \wedge \neg d$.

The failure of the replacement theorem may appear as sheer logical nonsense. In the Polish school traditionally only logics in which this property is valid are considered as real logics. Following Wójcicki such logics are called self-extensional (see [25]), rightly emphasizing the significance of the replacement property. Now it seems that the reason why the Poles like this property is not because they are against intensional logic, but because it allows easy algebraization using Tarski-Lindenbaum methodology.

As it has been argued elsewhere, it seems logical to consider that a logic is intensional if it does not obey the replacement theorem, to qualify as intensional self-extensional modal logics such as $\mathcal{S}5$, $\mathcal{S}4$, etc. seems quite absurd (see [6]). But a logic which is not self-extensional is not necessarily intensional. Here again anti-classical logic is a good example: replacement is not valid, but it cannot really be considered as intensional. In classical logic and the standard modal logics, due to the replacement theorem, it is the same to say that a proposition is logically equivalent to itself and to say it is logically equivalent to a very different proposition. This extensional feature turns many fundamental mathematical theorems trivial. But anti-classical logic is not less trivial: a proposition is never logically equivalent to itself but can be equivalent to its negation.

4 Being highly logical

It is not seriously possible to argue that anti-classical logic is not logical because features such as substitution and replacement do not hold. On the contrary, one may argue that logics such as classical logic $\mathcal{K}$ or a modal logic $\mathcal{S}5$ are not logical because they have these features. By arguing in this direction one may just want to say that $\mathcal{K}$ and $\mathcal{S}5$ do not properly describe our natural way of thinking. We surely don't want to claim that anti-classical logic $\overline{\mathcal{K}}$ is a good description of the way we naturally think, but it is not much more absurd than $\mathcal{K}$ or $\mathcal{S}5$. The structure $\overline{\mathcal{K}}$ is a useful tool to study logicality in the same way that $\mathcal{K}$ and $\mathcal{S}5$ are useful tools to develop, extend and challenge our concept of logicality. Considering that being logical does not reduce to blindly follow some rules, some laws of thought, but also questioning logicality and creating new rules, we can claim that to develop anti-classical logic is to be highly logical.

BIBLIOGRAPHY

[1] H. Bensusan, A. Costa-Leite and E. Gonçalves de Souza, "Logics and their galaxies". In *The Road to Universal Logic. Festschrift for the 50th Birthday of Jean-Yves Béziau*, Volume II, edited by A. Koslow and A. Buchbaum, Basel: Birkhaüser, 2015, pp. 243–255.
[2] J.-Y. Béziau, "Universal Logic", in *Proceedings of the 8th International Colloquium — Logica'94*, edited by T. Childers and O. Majer, Czech Academy of Sciences, Prague, 2004, pp. 73–93.
[3] J.-Y. Béziau, *Recherches sur la Logique Universelle*, PhD Thesis, Department of Mathematics, University Denis Diderot, Paris, 1995.
[4] J.-Y. Béziau, "What is paraconsistent logic ?", in *Frontiers of paraconsistent logic*, Research Studies Press, Baldock, 2000, pp. 95–112.
[5] J.-Y. Béziau, "Are paraconsistent negations negations?", in *Paraconsistency: the logical way to the inconsistent*, edited by W. Carnielli et al, Marcel Dekker, New-York, 2002, pp. 465–486.
[6] J.-Y. Béziau, "The philosophical import of Polish logic", in *Methodology and philosophy of science at Warsaw University*, edited by M. Talasiewicz, Semper, Warsaw, 2002, pp. 109–124.
[7] J.-Y. Béziau (editor), *Logica Universalis — Towards a general theory of logic*, Birkhäuser, Basel, 2005, 2nd edition, 2007.
[8] J.-Y. Béziau, *Les axiomes de Tarski*, in *La logique en Pologne*, edited by M. Rebuschi and R. Pouivet, Vrin, Paris, 2006, pp. 135–149.
[9] J.-Y. Béziau, "13 Questions about universal logic", *Bulletin of the Section of Logic*, vol. **35**, 2006, pp. 133–150.
[10] J.-Y. Béziau, "What is a logic? Towards axiomatic emptiness", *Logical Investigations*, vol. **16**, 2010, pp. 272–279.
[11] J.-Y. Béziau, "Logic is not logic", *Abstracta*, **6** (2010), 73–102.
[12] J.-Y. Béziau, W.A. Carnielli and D.M. Gabbay (editors), *Handbook of Paraconsistency*, College Publication, London, 2007.
[13] J.-Y. Béziau (editor), *Universal logic: an Anthology — From Paul Hertz to Dov Gabbay*, Birkhäuser, Basel, 2012.
[14] J.-Y. Béziau, "The power of the hexagon", *Logica Universalis*, vol. **6**, 2012, pp. 1–43.
[15] G. Birkhoff, "Universal algebra", *Comptes Rendus du Premier Congrès Canadien de Mathématiques*, University of Toronto Press, Toronto, 1946, pp.310–326.
[16] G. Birkhoff, "Universal algebra", in *Selected papers on algebra and topology by Garrett Birkhoff*, edited by G.-C. Rota and J.S. Oliveira, Birkhäuser, Basel, 1987.
[17] R. Descartes, *Discours de la méthode (Discourse on the method)*, Leyde, 1637.
[18] B. Pascal, *De l'esprit géométrique et de lart de persuader (The art of persuasion)*, 1657.
[19] J. Łoś and R. Suszko, "Remarks on sentential logics", *Indigationes Mathematicae*, vol. **20**, 1958, pp. 177–183.
[20] T. Skura, "A refutation theory", *Logica Universalis*, vol. **3**, 2009, pp. 293–302.
[21] T. Skura, "On refutation rules", *Logica Universalis*, vol. **5**, 2011, pp. 249–254.
[22] J.J. Sylvester, "Lectures on the Principles of Universal Algebra", *American Journal of Mathematics*, vol. **6**, 1884, pp. 270–286.
[23] A. Tarski, "Remarques sur les notions fondamentales de la méthodologie des mathématiques", *Annales de la Société Polonaise de Mathématiques*, vol. **6**, 1928, pp. 270–271.
[24] A.N. Whitehead, 1898, *A treatise on universal algebra*, Cambridge University Press, Cambridge.
[25] R. Wójcicki, *Theory of logical calculi*, Kluwer, Dordrecht, 1988.

[26] U. Wybraniec-Skardowska and J. Waldmajer, "On Pairs of Dual Consequence Operations", *Logica Universalis*, vol. **5**, 2011, pp. 177–203.
[27] J. Zygmunt, "Structural consequence operations and logical matrices adequate for them", in *Universal logic: an Anthology — From Paul Hertz to Dov Gabbay*, edited by J.-Y. Béziau, Birkhäuser, Basel, 2012, pp. 163–176.
[28] J. Zygmunt, "Tarski's first published contribution to general metamathematics", in *Universal logic: an Anthology — From Paul Hertz to Dov Gabbay*, edited by J.-Y. Béziau, Birkhäuser, Basel, 2012, pp.59–67.

Jean-Yves BEZIAU
Institute of Philosophy, University of Brazil, Rio de Janiero - UFRJ
Brazilian Academy of Philosophy - ABF
Brazilian Research Council - CNPq

Arthur BUCHSBAUM
Department of Informatics and Statistics
Federal University of Santa Catarina, Florianópolis, Brazil

Les Oppositions Modales dans la Logique d'Al Fārābī

SALOUA CHATTI

1. Introduction

Al Fārābī (873?-950)[1] est l'un des premiers logiciens et philosophes arabes. Baptisé le 'Second Maître' (le 'Premier Maître' étant Aristote), il était tenu en très haute estime par Avicenne et a également eu beaucoup d'influence sur Averroès, qui, notamment en matière de logique, défendait des positions très proches des siennes. Très inspiré par la philosophie grecque, notamment en philosophie politique, il a essayé de concilier les vues platonicienne et aristotélicienne. En matière de logique, il était influencé par Alexandre d'Aphrodise, qu'il citait souvent et dont l'interprétation de l'oeuvre d'Aristote lui paraissait la plus juste. Son oeuvre logique se compose de paraphrases d'Aristote, mais également de traités courts où il défend des positions plus personnelles, qu'il présente avec beaucoup de concision. Dans cet essai, je me propose de traiter non pas de l'ensemble de la logique d'Al Fārābī, ce qui exigerait beaucoup plus qu'un simple article, mais d'un point précis, à savoir les oppositions modales telles qu'elles sont présentées dans les deux traités où elles sont explicitement étudiées, qui sont intitulés respectivement: *Sharh Al 'Ibāra (Paraphrase de De l'interprétation)* et *Al Qawl fi Al 'Ibāra (Discours sur De l'interprétation)*. Mon but est de déterminer la position d'Al Fārābī sur ce sujet, ce qu'il retient d'Aristote, mais aussi son apport propre en cette matière. Les questions que je pose sont les suivantes: Comment Al Fārābī définit-il les notions de nécessité et de possibilité? Comment caractérise-t-il les oppositions modales? Y a-t-il des différences sur ce sujet entre le premier et le deuxième texte d'une part et entre sa (ou ses) position(s) et celle d'Aristote de l'autre?

Je commencerai par analyser les théories présentées dans chacun des deux textes, et montrerai ensuite qu'Al Fārābī apporte une contribution tout à fait originale dans la mesure où on trouve chez lui une première représentation incomplète de l'octogone de Buridan à laquelle sont ajoutés les sommets

---

[1] La date de naissance d'Al Fārābī est inconnue, mais les historiens la situent entre 870 et 873.

correspondant aux assertoriques et aux singulières. Sa théorie souffre toutefois des mêmes ambiguïtés que celle d'Aristote.

2. Les oppositions modales dans *Sharh Al 'Ibāra*

Commençons par introduire les deux textes où il est expressément question des modalités. Ces deux textes sont deux versions de *De l'interprétation*, l'une courte où Al Fārābi présente ses idées sans commenter Aristote et qui est intitulée *Al Qawl fi Al 'Ibāra*, l'autre beaucoup plus longue consistant en un commentaire très détaillé du texte aristotélicien avec citations à l'appui. Cette dernière version est intitulée *Sharh Al 'Ibāra* (*Paraphrase de De l'interprétation*). Soulignons à ce sujet que le texte correspondant aux *Premiers Analytiques* ne contient pas, contrairement à celui d'Aristote, de chapitres traitant du syllogisme modal. Mais cette absence peut être simplement liée à la perte de ces textes[2] car dans *Sharh Al 'Ibāra*, Al Fārābi renvoie souvent aux chapitres des *Premiers Analytiques* consacrés aux modalités. Pour cette raison, je me baserai uniquement sur les deux textes correspondant à *De l'interprétation*.

Dans *Sharh Al 'Ibāra*, Al Fārābi reprend et commente le texte même d'Aristote en suivant son argumentation. Aristote distingue en effet dans *De l'interprétation* entre quatre ordres de modalités qui sont: *Nécessaire que cela soit*, *Impossible que cela soit*, *Possible que cela soit* (= *Contingent que cela soit*) et *Non nécessaire que cela soit*. Le nécessaire correspond à l'universel et le possible au particulier; par suite les relations logiques entre ces propositions seront représentées par un carré où 'Nécessaire que cela soit' prend la place de **A**, 'Possible que cela soit' celle de **I**, 'Non nécessaire que cela soit' (ou 'possible que cela ne soit pas') celle de **O** et 'Impossible que cela soit' (= 'nécessaire que cela ne soit pas') celle de **E**. On aura donc deux couples de contradictoires (ni vraies ni fausses ensemble) qui sont *Nécessaire que cela soit* / *Non nécessaire que cela soit* d'une part et *Possible que cela soit* / *Pas possible que cela soit* (ou *impossible que cela soit*) de l'autre, deux propositions contraires (possiblement fausses ensemble mais jamais vraies ensemble), qui sont *Nécessaire que cela soit* et *Nécessaire que cela ne soit pas*, deux propositions subcontraires (possiblement vraies ensemble mais jamais fausses ensemble) qui sont: *Possible que cela soit* et *Possible que cela ne soit pas*. Enfin la subalternation (ou implication) a lieu entre *Nécessaire que cela soit* et *Possible que cela soit* d'une part et entre *Nécessaire que cela ne soit pas* et *Possible que cela ne soit pas* d'autre part.

---

[2]Joep Lameer dans [14, p. 59] évoque la perte de certains textes puisqu'il dit: "As stated earlier, al Fārābi takes no real interest in modal syllogistics although he will certainly have addressed this issue, in the part of his commentary on the *Prior Analytics* that has been *lost*", (mes italiques). Il ajoute que la syllogistique modale est évoquée dans son commentaire de *De l'interprétation*.

Al Fārābī reprend ces distinctions et ces relations et classe les propositions contradictoires comme suit:

"Nécessaire que cela soit / Pas nécessaire que cela soit
Pas possible que cela ne soit pas / Possible que cela ne soit pas
Impossible que cela ne soit pas / Pas impossible que cela ne soit pas

Nécessaire que cela ne soit pas / Pas nécessaire que cela ne soit pas
Pas possible que cela soit / Possible que cela soit
Impossible que cela soit / Pas impossible que cela soit"
[1, p. 217]

Comme on le voit, ce tableau représente les quatre ordres aristotéliciens puisque les trois lignes du haut à gauche, i.e. *Nécessaire que cela soit = Pas possible que cela ne soit pas = Impossible que cela ne soit pas*, représentent l'ordre du nécessaire [□a = ¬◊¬a]. Celles du bas à gauche, i.e. *Nécessaire que cela ne soit pas = Pas possible que cela soit = Impossible que cela soit*, représentent celui de l'impossible [□¬a = ¬◊a]. Celles du haut à droite, i.e. *Possible que cela ne soit pas = Non nécessaire que cela soit = Non impossible que cela ne soit pas*, représentent l'ordre du non nécessaire [◊¬a = ¬□a], et enfin celles du bas à droite, i.e. *Possible que cela soit = Non nécessaire que cela ne soit pas = Non impossible que cela soit*, représentent l'ordre du possible [◊a = ¬□¬a] [1, p. 217]. Ces correspondances montrent clairement que, comme Aristote, Al Fārābī admet les lois de dualité des opérateurs modaux puisque □¬ = ¬◊, par suite, ¬□¬ = ¬¬◊ = ◊; de même, ◊¬ = ¬□, par suite, ¬◊¬ = ¬¬□ = □.[3]

On peut cependant noter que, contrairement à Aristote, Al Fārābī a éliminé de ce tableau le contingent et a commencé par les propositions nécessaires. L'ordre adopté n'est pas le même que celui d'Aristote dans *De l'interprétation*, mais correspond davantage à ce qui est dit dans les *Premiers Analytiques* auquel Al Fārābī renvoie explicitement.

Les modalités du carré classique chez Aristote sont *de dicto* car elles portent sur tout le *dictum*, mais elles peuvent aussi s'appliquer aux modalités *de re* des propositions singulières comme le souligne Knuutila [13, §3], c'est-à-dire à des modalités incluses à l'intérieur des propositions et portant directement sur le prédicat. Bien qu'Al Fārābī ne distingue pas explicitement entre les modalités *de dicto* et les modalités *de re*, ses exemples concernent des indéfinies où les modalités sont *de re* et les indéfinies, pour lui, sont comparables aux singulières sur ce plan car leurs contradictoires se forment par la négation de la modalité. Soulignons à ce sujet que la distinction

---

[3]Nous avons introduit les symboles de la nécessité et de la possibilité utilisés par les modernes pour plus de simplicité et de lisibilité.

proprement médiévale *de re* / *de dicto* ou plus exactement *in sensu diviso* / *in sensu composito* [4] est plus complexe que ce que j'évoque ici. Car ce dont il est question chez Al Fārābi et plus tard chez Avicenne, qui en parle explicitement, est une simple distinction de portée de l'opérateur modal: dans un cas, cet opérateur est à l'intérieur de la proposition, dans l'autre, il est placé à l'extérieur et domine la proposition entière.

Comme Aristote, Al Fārābi considère que les propositions contraires sont d'abord les propositions nécessaire et impossible car "nécessaire qu'il ne soit pas est le contraire de nécessaire qu'il soit" [1, p. 199]. Par exemple "L'homme est nécessairement juste" et "L'homme nécessairement n'est pas juste" ne peuvent pas être vraies à la fois [1, p. 192]: elles sont en effet fausses toutes les deux et c'est pour cette raison qu'elles sont contraires.

Les propositions subcontraires sont les propositions possible et non nécessaire, car elles peuvent être vraies ensemble [1, p. 191] mais ne sont jamais fausses à la fois, car "possible que cela soit et possible que cela ne soit pas sont vrais du même sujet" puisque "ce qui voit maintenant, peut aussi ne pas voir" [1, p. 191][5].

Quant à la subalternation, elle s'applique au possible unilatéral mais pas au possible bilatéral comme le souligne Al Fārābi dans son commentaire du passage 13, 22b, 10-28, où Aristote pose explicitement la question de l'implication du possible par le nécessaire. Il ressort de ce commentaire que quand le possible est unilatéral, "il est vrai du nécessaire" [1, p. 206], ce qui signifie que "Nécessaire que cela soit" implique "Possible que cela soit", par contre le possible bilatéral i.e. "possible que cela soit et possible que cela ne soit pas", est contraire à l'impossible et au nécessaire comme le souligne Al Fārābi dans ce qui suit: "Quant à 'nécessaire que cela soit' et 'nécessaire que cela ne soit pas', aucun d'eux ne peut être vrai en même temps que la conjonction des deux autres, qui est 'possible que cela soit et que cela ne soit pas'. Car quand 'nécessaire que cela soit' est vrai, 'possible que cela ne soit pas' n'est pas vrai et quand 'nécessaire que cela ne soit pas' est vrai 'possible que cela soit' n'est pas vrai" [1, p. 204]. La subalternation ne concerne donc pas ce type de possibilité, elle ne vaut que pour le possible unilatéral. Elle concerne aussi le non nécessaire, dont on peut dire qu'il est

---

[4]Globalement la modalité *in sensu composito* correspond à la modalité *de dicto* et place l'opérateur modal au début de la proposition tandis que les modalités *in sensu diviso* placent la modalité à l'intérieur de la proposition et correspondent à la modalité *de re*. Mais comme le souligne Knuutila [13, §3], la distinction fait aussi intervenir des considérations temporelles comme en témoigne l'exemple qu'il donne dans son article et qui est le suivant: "A standing man can sit". Interprétée *de dicto*, cette phrase dit: "Il est possible qu'un homme s'assoie et soit debout en même temps" et dans ce cas elle est fausse, mais interprétée *de re*, elle dit simplement: "Un homme qui est maintenant debout peut s'asseoir" et dans ce cas, elle est vraie.

[5]Toutes les traductions des textes d'Al Fārābi sont les miennes.

impliqué par l'impossible. Cette implication est exactement parallèle à la précédente et peut être justifiée de la même manière.

Les deux sens du possible sont évoqués explicitement à la page 206 où Al Fārābī dit: "Si on considère le possible d'une certaine manière, il est vrai du nécessaire, mais si on le considère d'une autre manière, il n'est pas compatible avec lui. Le possible doit donc être pris dans deux acceptions" [1, p. 206]. Ces deux acceptions sont précisément le possible unilatéral et le possible bilatéral. Il ressort de la lecture qu'Al Fārābī fait de *De l'interprétation* que le triangle des contraires est présent même au niveau de ce traité[6] et se superpose au carré des oppositions modales.

Le possible bilatéral est considéré par Al Fārābī comme le sens authentique du possible, car c'est lui qui exprime la spécificité du possible et le distingue du nécessaire. Il exprime ce qu'on peut appeler le sens étroit du possible qui se distingue d'un sens large englobant plusieurs acceptions. Pour bien marquer la différence entre les sens étroit et large du possible, Al Fārābī précise que le sens large englobe le nécessaire tandis que le sens étroit le dépasse. Le nécessaire lui-même, précise-t-il, peut avoir une acception large et une acception étroite. En effet le nécessaire est: 1) "ce qui n'a pas cessé d'être et ne cessera pas d'être", 2) ce qui est "nécessaire tant que son sujet existe" et 3) "ce dont l'existence est nécessaire tant que lui-même existe" [1, p. 214]. Le possible, d'après lui, possède ces trois acceptions du nécessaire, mais aussi une quatrième acception, qui le distingue du nécessaire et le rend plus général. Cette quatrième acception est précisément le possible bilatéral, c'est-à-dire "ce qui peut être et peut ne pas être dans le futur" [1, p. 214]. Commentant le texte aristotélicien, Al Fārābī précise que "le possible est plus général que le nécessaire car il peut être dit du nécessaire et de ce qui n'est pas nécessaire" [1, p. 214]. Au sens large, le possible englobe donc le nécessaire, car ce qui est nécessaire est forcément possible puisque s'il ne l'était pas il serait impossible, ce qui est absurde, mais il le dépasse par son sens étroit qui admet les contraires. Quant au sens étroit du nécessaire c'est la nécessité absolue, c'est-à-dire le premier des trois sens précités. Son sens large comprend les trois sens précités où les deux derniers correspondent à l'assertorique et à l'existentiel. La nécessité de ces derniers est liée à l'existence car elle peut être comprise comme étant ce qui est nécessaire au moment précis où il est et non d'une façon permanente.

---

[6]La position aristotélicienne est considérée par les logiciens modernes en particulier comme à la fois complexe et ambiguë. Ainsi Blanché considère qu'Aristote emploie le "possible-contingent" tantôt en un sens qui "flotte" entre "le pur possible et le possible bilatéral", tantôt au sens unilatéral, tantôt au sens bilatéral [9, p. 68]. Toutefois, selon Blanché, le sens bilatéral semble privilégié dans les *Premiers Analytiques*, et le sens unilatéral est privilégié dans *De l'interprétation* [9, p. 72].

Le possible bilatéral est lié principalement aux propositions parlant du futur. Le futur est ouvert à tous les possibles et ne doit pas, selon Al Fārābi, être déterminé comme le pensent les mégariques. En adoptant cette position, Al Fārābi défend donc explicitement la thèse d'Aristote contre celle des mégariques. Toutefois, la négation du possible bilatéral n'est pas donnée par Aristote. Pour justifier cette absence, Al Fārābi considère que la négation du possible, dans tous les sens qu'il peut prendre, est l'impossible, car seul l'impossible peut nier aussi bien le sens large que le sens étroit du possible. La figure correspondant à ces distinctions semble être un hexagone tronqué, contenant un carré et un triangle des contraires[7]. On voit donc que la position défendue dans ce premier texte est une réplique de celle d'Aristote. Toutefois on peut constater que dès ce premier texte Al Fārābi introduit les modalités *de re*, comme en témoignent ses exemples personnels où les termes modaux sont contenus à l'intérieur des propositions. Qu'en est-il du deuxième traité? C'est ce que nous verrons dans ce qui suit.

3. Les modalités dans *Al Qawl fi Al 'Ibāra*

Dans ce traité, Al Fārābi analyse les modalités en considérant tous les types de propositions: singulières, indéfinies et quantifiées. Le texte lui-même n'est pas à proprement parler un commentaire des traités aristotéliciens. Al Fārābi parle en son nom propre, bien qu'il renvoie parfois à Aristote et à ses commentateurs comme Alexandre d'Aphrodise, et il introduit des approfondissements qui conduisent au dépassement du carré classique des modalités. Cette analyse des modalités se fait d'abord par l'introduction des termes modaux à l'intérieur des propositions aussi bien singulières qu'indéfinies et quantifiées. Les modalités étudiées sont donc des modalités *de re* au sens précis que nous avons évoqué plus haut. Ces modalités affectent la façon dont le prédicat est attribué au sujet. Ainsi, dans les propositions singulières ou indéfinies, les propositions modales sont exprimées ainsi: (1) "Zayd peut marcher" ou encore (2) "L'homme peut être juste" [2, pp. 105-106]. Dans ces propositions, la négation peut porter soit sur la modalité, soit sur le prédicat soit sur la copule, mais la contradictoire est la proposition où la négation porte sur la modalité. Ainsi, la contradictoire de (1) est "Zayd ne peut pas marcher", celle de (2) est "L'homme ne peut pas être juste". Quand la négation porte sur la copule ou le verbe principal, on obtient des propositions négatives mais pas contradictoires aux propositions initiales, comme par exemple "Zayd possiblement ne marche pas" ou "L'homme possiblement n'est pas juste" [2, p. 106] (traductions littérales). Ces propositions sont appelées des "possibles négatives" [2, p. 110] et elles se distinguent des

---

[7]Pour une illustration de cet hexagone tronqué, voir par exemple Alessio Moretti [15, p. 89].

véritables négations des possibles qui expriment l'impossibilité.

Les oppositions préalablement admises pour les propositions catégoriques quantifiées, indéfinies et singulières sont la contradiction, la contrariété et la subcontrariété. Les indéfinies assertoriques, bien que non quantifiées, sont traitées comme des particulières, car elles sont subcontraires [3, p. 122]. Ces oppositions seront reprises et appliquées aux propositions contenant des modalités explicites. Al Fārābī procède d'une manière systématique en faisant porter la négation sur la copule, la modalité, le prédicat ou le quantificateur.

Par cette analyse, il détermine les contradictoires des propositions singulières, quel que soit leur *dictum* (ou contenu). Ces propositions et leurs contradictoires sont illustrées par les exemples suivants:

(1) "Zayd peut marcher" / "Zayd ne peut pas marcher"
(2) "Zayd possiblement ne marche pas" / "Zayd non possiblement ne marche pas"
(3) "Zayd possiblement n'est pas non-savant" / "Zayd non possiblement n'est pas non-savant"[8] [2, p. 106-107].

Formalisées, elles deviennent: (1) Z est $\Diamond A$ / Z est $\neg \Diamond A$, (2) Z est $\Diamond \neg A$ / Z est $\neg \Diamond \neg A$, (3) Z est $\Diamond \neg \neg A$ / Z est $\neg \Diamond \neg \neg A$.

En utilisant les équivalences mentionnées dans le premier traité, qu'on peut considérer comme admises par Al Fārābī, on peut reformuler plus simplement certaines de ces propositions et obtenir les propositions suivantes:

"Zayd non nécessairement marche" / "Zayd nécessairement marche" [Z est $\neg \Box A$/ Z est $\Box A$ = (2) ci-dessus].

"Zayd est possiblement savant" / "Zayd est nécessairement non savant" [Z est $\Diamond A$ / Z est $\Box \neg A$ = (3) ci-dessus]

Par ailleurs la "possible négative" et sa correspondante affirmative comme par exemple, "Zayd possiblement n'est pas savant" [2, p. 106] et "Zayd est possiblement savant" peuvent être toutes deux vraies. Leur relation est donc plutôt la subcontrariété, comme dans le traité précédent.

Par contre, la "nécessaire négative" est la contraire de sa correspondante affirmative. Ainsi par exemple: "Zayd ne peut pas être non savant" [2, p. 106] (= "Zayd est nécessairement savant"), est la contraire de "Zayd ne peut pas être savant" [2, p. 107] (= "Zayd nécessairement n'est pas savant"), puisqu'elles sont toutes les deux fausses.

---

[8]Malgré le caractère alambiqué de ces formulations, nous avons préféré les garder car ce sont les traductions littérales du texte d'Al Fārābī, lequel considère systématiquement toutes les portées possibles de la négation dans ses exemples. Soulignons que même en arabe, elles ne sont pas très naturelles; le souci d'Al Fārābī est donc avant tout logique.

Quant aux propositions quantifiées contenant des termes modaux explicites, Al Fārābī affirme que leurs contradictoires peuvent être formées en faisant porter la négation sur le quantificateur [2, pp. 105-106], ce qui indique que la modalité est interne à la proposition. Comme nous le verrons par la suite, cette façon de nier les propositions modales quantifiées conduit bien aux contradictoires car le changement d'opérateur modal se fait dans le même temps et sous l'effet de la négation. En effet, dans son analyse des propositions quantifiées, Al Fārābī considère les universelles possibles dont le *dictum* est affirmatif ou négatif, mais en les niant, il obtient les particulières correspondantes, ce qui complète son analyse, et il précise également qu'on peut procéder de la même façon pour la modalité du nécessaire. Il montre ainsi ce que sont les contradictoires et les contraires de chacune de ces propositions. Ainsi, la proposition:

"Tout homme possiblement marche" [2, p. 106] [= Tout A est possiblement B = $(\forall x)(Ax \supset \Diamond Bx)$] (1)

a une contradictoire qui est: "Non tout homme possiblement marche" [2, p. 106] [= Non tout A est possiblement B = $\neg(\forall x)(Ax \supset \Diamond Bx)$ = Quelques A ne sont pas possiblement B = $(\exists x)(Ax \wedge \neg \Diamond Bx)$ = Quelques A sont nécessairement non B = $(\exists x)(Ax \wedge \Box \neg Bx)$] (2)

et une contraire qui est: "Aucun homme possiblement ne marche" [2, p. 106] [= Aucun A n'est possiblement B = $(\forall x)(Ax \supset \neg \Diamond Bx)$ = Tout A est nécessairement non B = $(\forall x)(Ax \supset \Box \neg Bx)$] (3)

Quand le *dictum* est négatif comme dans la proposition suivante:

"Tout homme possiblement ne marche pas" [2, p. 107] [= $(\forall x)(Ax \supset \Diamond \neg Bx)$] (4)

la contradictoire est: "Non tout homme possiblement ne marche pas" [2, p. 107] [= $\neg(\forall x)(Ax \supset \Diamond \neg Bx)$ = Quelques hommes non possiblement ne marchent pas = $(\exists x)(Ax \wedge \neg \Diamond \neg Bx)$ = Quelques hommes nécessairement marchent = $(\exists x)(Ax \wedge \Box Bx)$] (5)

et la contraire est: "Aucun homme possiblement ne marche pas" [2, p. 107] [= Aucun A n'est possiblement non B = $(\forall x)(Ax \supset \neg \Diamond \neg Bx)$ = $(\forall x)(Ax \supset \Box Bx)$] (6)

Si la proposition contient la modalité du nécessaire, comme la suivante: "Tout homme est nécessairement juste" [= $(\forall x)(Ax \supset \Box Bx)$] (6)

sa contradictoire est exprimée ainsi: "Non tout homme est nécessairement juste" [= $\neg(\forall x)(Ax \supset \Box Bx)$ = $(\exists x)(Ax \wedge \neg \Box Bx)$ = $(\exists x)(Ax \wedge \Diamond \neg Bx)$] (7)

et sa contraire est: "Aucun homme n'est nécessairement juste" [= $(\forall x)(Ax \supset \neg \Box Bx) = (\forall x) (Ax \supset \Diamond \neg Bx)$] (4).

De même, la proposition suivante:

"Tout homme est nécessairement non juste" [= $(\forall x)(Ax \supset \Box \neg Bx)$] (3)

a comme contradictoire: Non tout homme est nécessairement non juste [= $\neg(\forall x) (Ax \supset \Box \neg Bx) = (\exists x) (Ax \wedge \neg \Box \neg Bx) = (\exists x) (Ax \wedge \Diamond Bx)$] (8)

et comme contraire: Aucun homme n'est nécessairement non juste [= $(\forall x) (Ax \supset \neg \Box \neg Bx) = (\forall x) (Ax \supset \Diamond Bx)$] (1)

En éliminant les redondances, on arrive aux huit sommets de l'octogone de Buridan, qui sont respectivement les propositions quantifiées suivantes:

Tout A est nécessairement B (= (6) ci-dessus)
Tout A est possiblement B (= (1) ci-dessus)
Quelques A sont nécessairement B (= (5) ci-dessus)
Quelques A sont possiblement B (= (8) ci-dessus)
Tout A est nécessairement non B (= (3) ci-dessus)
Tout A est possiblement non B (= (4) ci-dessus)
Quelques A sont nécessairement non B (= (2) ci-dessus)
Quelques A sont possiblement non B (= (7) ci-dessus)

Si on inclut les singulières au sein des propositions modales opposées, on obtient la figure suivante:

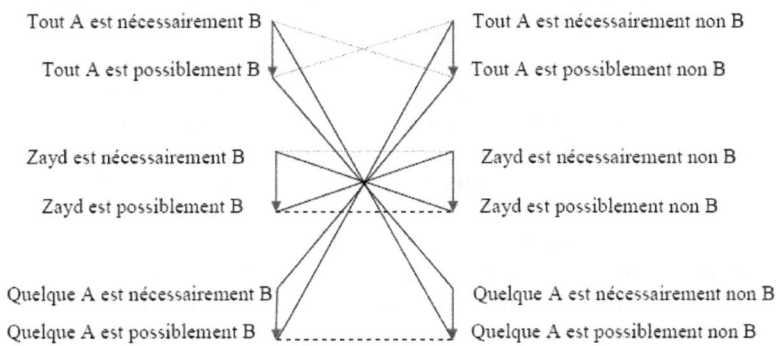

Figure 1.

où les lignes en pointillés traduisent la contrariété, les lignes grasses la contradiction, les tirets la subcontrariété, et les flèches la subalternation. Dans cette figure nous avons inclus la subcontrariété entre les deux particulières car l'indéfinie est comparable à une particulière quand elle est assertorique, les indéfinies affirmative et négative étant des subcontraires comme le précise Al Fārābi dans [3, p. 122]. Puisque les indéfinies possibles, affirmative et négative, sont subcontraires et particulières, on peut en déduire que les particulières possibles, négative et affirmative, sont elles aussi des subcontraires bien qu'Al Fārābi ne l'ait pas affirmé explicitement étant donné qu'il n'a évoqué, pour les propositions quantifiées, que les contraires et les contradictoires. Nous avons également inclus la subalternation entre le nécessaire et le possible unilatéral. Comme on l'a déjà vu, le nécessaire est analysé en termes d'existence et de permanence (notion temporelle) et le possible en termes de potentialité et d'absence de permanence.

Cette figure contient tous les sommets de l'octogone de Buridan et y ajoute quatre sommets représentant les propositions singulières qui forment un carré. Toutefois si le carré est complet, l'octogone, lui, n'est pas complet puisque les relations admises par Buridan ne sont pas toutes représentées dans la figure. En effet, les subcontraires des propositions modales quantifiées ne sont pas évoquées explicitement, et seule une contraire de "Tout A est nécessairement B" et *une* contraire de "Tout A est nécessairement non B" sont explicitement évoquées. Or ces propositions ont trois contraires chacune chez Buridan.[9] Par contre, la subcontrariété entre les propositions possibles singulières est bien présente chez lui, et la subalternation entre le nécessaire et le possible unilatéral est aussi présente et applicable à tous les types de propositions. Bien qu'elle ne soit pas considérée comme une opposition véritable, la subalternation entre le nécessaire et le possible unilatéral est en effet une relation admise par Al Fārābi.

Toutefois, juste après avoir analysé ces modalités *de re*, Al Fārābi présente une définition des modalités où il introduit l'assertorique. Il dit en effet "les modalités sont au nombre de trois: le nécessaire (*darūrī*), le possible (*mumkin*) et l'absolu ou assertorique (*motlaq*)" [2, p. 108]. Le nécessaire est "le permanent, ce qui a été, est encore et ne peut pas ne pas être, à quelque moment que ce soit", le possible est "ce qui n'est pas maintenant existant, mais peut être et ne pas être à n'importe quel moment du futur", et l'assertorique est "ce qui était de la nature du possible et est maintenant existant après qu'il ait été possible qu'il soit et qu'il ne soit pas, et il est possible aussi qu'il ne soit pas dans le futur" [2, p. 108]. La définition du nécessaire met l'accent sur le sens absolu du traité précédent, celle du possible met l'accent sur le possible bilatéral (mais à la fin du texte, Al Fārābi réaffirme

---

[9]Cf. par exemple Stephen Read [16, p. 13] et également Fabien Schang [18] où les mêmes relations sont présentées entre des propositions formalisées contenant des modalités et des quantificateurs.

l'ambiguïté du possible comme on le verra par la suite), et l'assertorique correspond à l'existentiel c'est-à-dire à ce qui est réel à un moment donné, mais n'est pas permanent. L'assertorique et l'existentiel sont ainsi "synonymes" [2, p. 109]. L'assertorique équivaut donc à la modalité nulle, ce qui est affirmé explicitement dans le passage suivant: "L'usage veut que la [proposition] assertorique (*motlaqa*) soit traduite par l'élimination de toutes les modalités et qu'elle n'exprime ni la possibilité ni la nécessité. Ils ont considéré que cette élimination est sa modalité à elle" [2, p. 109]. En disant cela, il renvoie à Alexandre d'Aphrodise qui est explicitement évoqué. On peut donc comprendre que cet usage date des explications d'Alexandre du texte d'Aristote. Or, c'est dans les *Premiers Analytiques* qu'Aristote combine les propositions assertoriques avec les propositions nécessaires et possibles, mais les propositions assertoriques (ou pures) sont précisément celles qui sont étudiées dans la syllogistique non modale comme en témoignent les intitulés des chapitres 2 et 3 de ce traité[10]. Cette introduction de l'assertorique révèle une divergence par rapport au texte précédent. Elle résulte d'une certaine interprétation des *Premiers Analytiques* qui peut être contestée, puisque dans ce traité, l'assertorique n'est pas forcément une modalité nouvelle ajoutée au nécessaire et au possible. Mais Al Fārābi y voit au contraire une interprétation plus juste de la pensée d'Aristote, se basant en cela sur l'interprétation qu'en donne Alexandre d'Aphrodise [2, p. 109].

Les modalités logiques dont il a été question jusqu'ici se distinguent des modalités matérielles qui sont liées au contenu de la proposition et à la nature du lien entre le sujet et le prédicat. En effet les propositions ne contenant pas de modalités explicites sont déjà modalisées au niveau de leur contenu. Car quand le lien entre le sujet et le prédicat est permanent, la proposition est nécessaire matériellement comme dans la phrase "Trois est impair", quand ce lien est contingent ou non permanent, la proposition est possible matériellement, comme dans la phrase "L'homme est blanc", quand le prédicat ne peut jamais être attribué au sujet, la proposition est impossible matériellement, comme dans la phrase "L'homme est une pierre". C'est en fonction de ces modalités matérielles que les relations d'opposition comme la contradiction, la contrariété et la subcontrariété sont définies. Ainsi, par exemple, la contrariété se caractérise par le fait que les deux propositions ne partagent pas la même valeur de vérité dans le nécessaire et l'impossible

---

[10]Respectivement: "§2: La conversion des propositions pures", et "§3: La conversion des propositions modales" [5, I, 1, 25a, et 25a, 25-30]. En réalité, ces intitulés ne sont pas ceux d'Aristote lui-même car ils figurent dans la traduction française mais pas, par exemple, dans la traduction anglaise. Il n'en demeure pas moins qu'Aristote sépare bien le chapitre où il traite des propositions modales (§3) de celui où il traite des propositions assertoriques (§2), ce qui montre qu'il regroupe les propositions assorties d'une modalité explicite et les distingue nettement de celles qui ne contiennent aucune modalité explicite.

matériels, mais sont fausses dans le possible, et la subcontrariété par le fait qu'elles n'ont pas la même valeur de vérité dans le nécessaire et l'impossible matériels mais sont vraies dans le possible [3, p. 122]. Ainsi les propositions particulières suivantes: "quelques hommes sont justes" et "quelques hommes ne sont pas justes" sont possibles matériellement et sont toutes les deux vraies. Par contre quand une proposition est impossible matériellement comme "quelques hommes sont des pierres" elle est fausse quand elle est affirmative et vraie quand elle est négative (quelle que soit sa quantité), et quand elle est nécessaire matériellement, elle est vraie quand elle est affirmative et fausse quand elle est négative (quelle que soit sa quantité). Par exemple "quelques hommes ne sont pas des animaux" et "aucun homme n'est un animal" sont toutes les deux fausses. Cette notion de modalité matérielle sera reprise par la suite par Avicenne et Averroès[11] entre autres et semble également courante chez les médiévaux, comme le souligne par exemple Knuutila dans [13, §3].

Les modalités logiques peuvent être introduites dans tous ces types de phrases, et la négation peut porter soit sur les termes modaux, soit sur le verbe principal de la phrase. Dans le premier cas, on obtient la contradictoire de la proposition, dans le deuxième cas, on obtient des propositions qui sont "possibles négatives" ou "nécessaires négatives". En effet, l'analyse et les reformulations précédentes ont montré que la contradictoire d'une proposition modale est finalement la proposition où la modalité elle-même est niée, même quand la proposition est quantifiée. Celles où les négations portent sur le verbe principal ou le prédicat sont soit les contraires soit les subcontraires de leurs correspondantes affirmatives.

La définition du possible dans ce deuxième traité aussi est complétée par la distinction entre les deux types de possibilité précédemment cités puisque Al Fārābī réaffirme à la fin de son texte l'ambiguïté du possible, comme on peut le constater dans le passage suivant: "Le possible se dit par communauté de nom (*bi-ishtirāk al-'ism*) de quatre sens, les trois premiers étant ce qui peut être dit du nécessaire, et de l'assertorique, et le quatrième étant ce qui n'est pas existant maintenant, mais dont on peut dire qu'il se peut qu'il soit et qu'il ne soit pas dans le futur. Toutefois, le véritable possible est celui qui correspond au quatrième de ces sens." [2, p. 114] Ce passage montre que sur le fond, la définition du possible n'a pas changé d'un traité à l'autre, puisque le possible est pris dans le deuxième traité aussi dans les deux sens: unilatéral ou compatible avec le nécessaire et le réel, et bilatéral ou incompatible avec le nécessaire et l'impossible. Le sens bilatéral est plus spécifiquement lié aux phrases du futur, et plus particulièrement aux singulières d'entre elles, et son adoption est due à des considérations

---

[11]Voir à ce sujet [Chatti, 2012].

philosophiques beaucoup plus que logiques. Ces considérations concernent notamment le problème du futur contingent dans sa relation avec l'action humaine, mais aussi avec l'action divine. La position d'Al Fārābī sur ce sujet se rapproche de celle d'Aristote. Il considère que les propositions contingentes du futur, même si elles sont contradictoires comme par exemple "Zayd ira demain au marché" et "Zayd n'ira pas demain au marché", ne le sont pas d'une manière stricte, car leur contradiction n'est pas strictement déterminée, dans la mesure où aucune d'elles n'est strictement vraie ni strictement fausse. Elles sont donc indéterminées en elles-mêmes et pour nous car non seulement personne ne peut dire laquelle est vraie et laquelle est fausse, mais en elles-mêmes, aucune n'est vraie ou fausse d'une manière déterminée [2, pp. 110-111]. Les propositions futures contingentes ne sont donc pas déterminées strictement; pour cette raison, le futur admet le possible bilatéral. Cette indétermination est justifiée également par des raisons théologiques car, dit-il, si le futur était déterminé, Dieu "ne pourrait pas rendre existant ce qui était inexistant, ou rendre inexistant ce qui était existant, à tout moment, ni au moment où Il le voudrait" [2, p. 113]. Décréter le caractère déterminé du futur équivaudrait alors à limiter la capacité divine à le choisir, ce qui est selon Al Fārābī, impensable.

Si on tient compte de ces définitions et de ces ajouts, la figure adoptée sera plus complexe dans la mesure où il faudra inclure les assertoriques et le possible bilatéral, qui semble être le sens authentique du possible. On aura donc une figure comportant dix neuf sommets en tout. Cette figure est la suivante (Figure 2):

On peut décomposer cette figure en trois parties qui sont: l'octogone des propositions modales quantifiées *de re* auquel sont ajoutés quatre sommets correspondant aux assertoriques, un hexagone comparable à l'hexagone de Czeżowski[12] – qui est une extension du carré classique des propositions quantifiées obtenu par l'ajout des propositions singulières à ces dernières – et un triangle des contraires. Toutefois, l'hexagone d'Al Fārābī concerne exclusivement les propositions singulières et n'est pas aussi complet que celui de Czeżowski. De même plusieurs autres relations manquent dans presque toutes les figures, comme par exemple la subalternation entre les propositions universelles et particulières qui n'est pas évoquée par Al Fārābī. Les trois figures précitées sont les figures 3, 4 et 5.

Ces trois figures illustrent la complexité de la théorie, mais elles montrent également que l'ambiguïté de la position aristotélicienne sur le possible est présente aussi dans la théorie d'Al Fārābī. La différence principale entre les

---

[12]Cet hexagone est présenté par Czeżowski dans [11]. Il est également évoqué par plusieurs auteurs; voir par exemple Fabien Schang [19]. Notons toutefois qu'il manque à l'hexagone d'Al Fārābī, un couple de contraires et un couple de subcontraires.

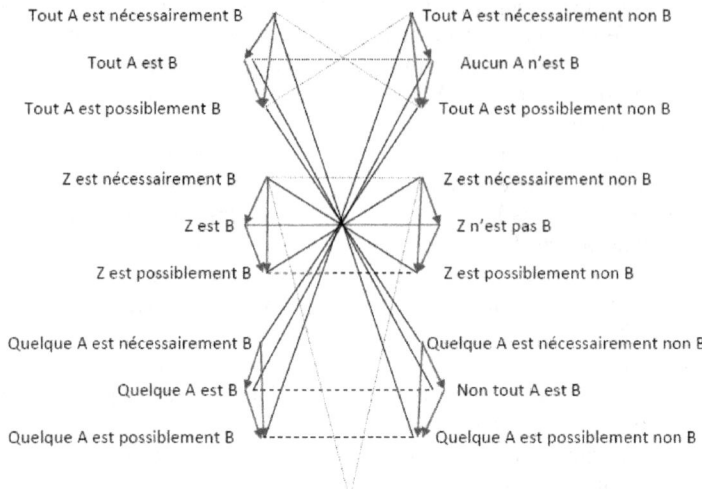

Figure 2.

deux auteurs semble concerner les assertoriques, car dans la mesure où Al Fārābī inclut l'assertorique à l'intérieur des modalités, il semble admettre l'axiome (**T**) $A \supset \Diamond A$ aussi bien que l'axiome (**M**) $\Box A \supset A$, qui sont ainsi ajoutés à l'axiome (**D**) $\Box A \supset \Diamond A$, lequel est admis par Aristote aussi. Ces axiomes s'appliquent aussi aux propositions quantifiées. Toutefois, bien que certains passages de son texte puissent faire penser qu'il évoque l'itération et le cumul des modalités, notamment quand il parle de propositions qui sont "nécessaires logiquement mais possibles matériellement" ou encore "nécessaires aussi bien matériellement que logiquement"[13] en illustrant ces cas par des exemples mathématiques ou autres, dans la mesure où les modalités matérielles ne sont pas traduites par des symboles spécifiques, il n'y a pas d'itération proprement dite. Pour cette raison, et si on considère le lien explicite entre les modalités aléthiques et les notions temporelles, il semble que le système défendu par Al Fārābī soit assez faible et proche du système **D** ou **T**.[14]

Passons maintenant à l'évaluation de cette théorie.

---

[13]Voir par exemple [2, p. 108].
[14]Nous nous basons dans ce classement des systèmes modaux sur [12].

(1)

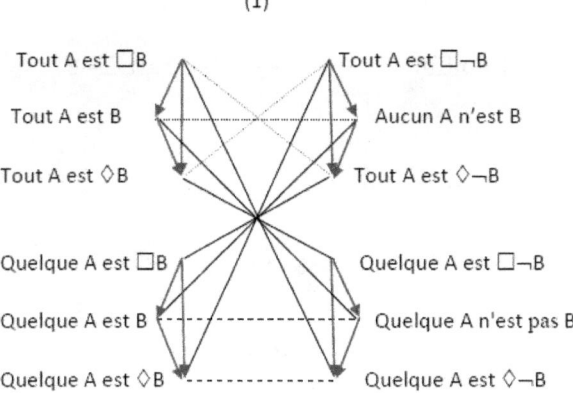

Figure 3.

(2)

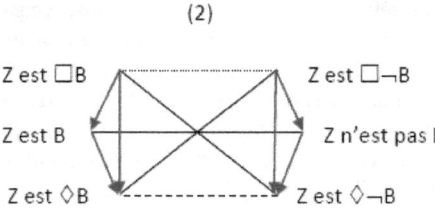

Figure 4.

4. Evaluation de la théorie

La première chose à souligner, c'est l'apport d'Al Fārābī au sujet des distinctions modales qui concernent les propositions quantifiées. D'après notre analyse, il a distingué, avant Buridan, les oppositions entre les propositions quantifiées, quoique d'une manière incomplète, et a donné tous les sommets de l'octogone. De même, on peut déceler dans ses distinctions concernant les oppositions des propositions singulières un hexagone qui rappelle celui de Czeżowski. Dans cette mesure, sa théorie est plus précise que celle d'Aristote. Toutefois, on peut s'interroger sur sa cohérence, car on constate des différences entre les deux traités dans la mesure où le second traité élargit la notion de modalité en introduisant l'assertorique, ce qui a pour conséquence que les positions défendues ne sont pas exactement les mêmes.

26   Saloua Chatti

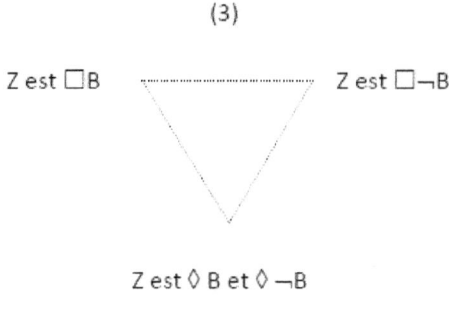

Figure 5.

Mais on peut atténuer cette différence en constatant d'une part que dès le premier traité, l'assertorique est considéré comme une forme de nécessité liée à l'existence, ce qui plaide pour son assimilation à une modalité, d'autre part que la définition du possible, qui conserve ses deux sens et ses liens avec le nécessaire dans le deuxième traité est la même d'un traité à l'autre. Il y a donc malgré les différences une certaine continuité entre les deux traités. Toutefois, la théorie du possible défendue est aussi ambiguë que celle d'Aristote, qui défend lui aussi deux visions du possible sans les distinguer clairement, comme l'ont montré ses commentateurs et comme on l'a signalé ici même (Cf. note 6, plus haut) car bien que le possible unilatéral et bilatéral soient clairement distingués au niveau de la proposition affirmative, la négation du possible bilatéral n'est pas séparée de celle du possible unilatéral, puisque c'est aussi l'impossibilité. Ceci introduit une forme d'incohérence dans la théorie puisque l'impossible est le contraire du possible bilatéral mais est aussi dit nier tous les types du possible: il est donc à la fois le contraire et le contradictoire du possible bilatéral. Al Fārābī, pas plus qu'Aristote, ne donne pas la négation spécifique du possible bilatéral, ce qui rend la théorie asymétrique. Or cette négation du possible bilatéral, ou ce qu'on appelle aujourd'hui le non contingent, existe et elle est donnée très clairement par Avicenne qui affirme, bien avant Blanché,[15] ce qui suit: "Ce qui n'est pas possible n'est pas l'impossible (*mumtana'*), il est plutôt le nécessaire soit dans l'existence soit dans l'inexistence" [6, p. 117], ce qui signifie que le non contingent est la proposition disjonctive "nécessaire que cela soit ou nécessaire que cela ne soit pas". Cette proposition disjonctive est bien la négation du contingent chez les modernes. Avicenne est donc

---
[15]Blanché la donne notamment dans [7] et [8, p. 79].

plus clair qu'Al Fārābī sur ce sujet, car il sépare les deux types de possibilité d'une façon plus nette que ses prédécesseurs. Par ailleurs, comme on l'a souligné, les modalités des propositions quantifiées sont internes et supposent pour cette raison ce que les médiévaux appellent l'ampliation du sujet qu'on retrouve chez Buridan, mais cette ampliation, comme le signalent Stephen Read et Wilfrid Hodges, a fait l'objet de critiques de la part de Guillaume d'Ockham[16] qui considère que la proposition particulière possible peut être lue de deux manières et que seule l'une de ces lectures en fait la contradictoire de la nécessaire universelle qui, elle, n'est pas ambiguë. En effet "Quelques A sont possiblement non B" est chez Buridan mais aussi Al Fārābī la contradictoire de "Tout A est nécessairement B"; or Stephen Read et Wilfrid Hodges affirment que d'après Guillame d'Ockham, cette proposition possible négative peut être interprétée de deux manières: 1) "Quelque chose qui est A est possiblement non B" ou 2) "Quelque chose qui peut être A est possiblement non B" [Cf. [16], p. 910 et [17], p. 16, mes italiques]. De ces deux lectures, seule la première est la contradictoire de la proposition nécessaire, qui, elle, n'admet qu'une seule interprétation. La même critique vaut pour la nécessaire négative et sa contradictoire possible affirmative. Cela montre que l'interprétation *de re* des propositions modales, admise par Buridan mais également par Al Fārābī, contient en elle-même certaines ambiguïtés qui peuvent rejaillir sur les oppositions. Elle doit donc recevoir un surcroît de précision. Soulignons là encore que cette interprétation *de re* a été distinguée de l'interprétation *de dicto* des modalités quantifiées par Avicenne aussi dans [6] où on lit: "Si on dit 'Tout homme est possiblement écrivain', ceci est naturel, et signifie: chacun des hommes peut être écrivain, mais ... quand on dit: 'Il est possible que tous les hommes soient des écrivains', où le possible est la modalité de l'universelle et du quantificateur, cela peut paraitre douteux" [6, p. 115, ma traduction]. Dans ce passage, Avicenne évoque la différence de portée de la modalité, mais également ce qui en résulte au niveau de la valeur de vérité de la proposition. Les deux propositions n'ont manifestement pas la même valeur de vérité puisque si la première est clairement vraie, la deuxième est probablement fausse.

5. Conclusion

En conclusion, on peut dire que la théorie d'Al Fārābī sur les oppositions modales contient des distinctions réellement nouvelles par rapport à celles d'Aristote, distinctions qui préfigurent notamment celles de Buridan. Les propositions singulières sont explicitement étudiées et l'assertorique est con-

---

[16]Le texte d'Ockham sur lequel se base S. Read, par exemple, est: *Summa Logicae* (1974), Franciscan Institute Publications, Edition P. Boehner et al.

sidéré comme une modalité, ce qui conduit à un hexagone des propositions singulières et à une figure plus complexe incluant l'octogone de Buridan pour les propositions quantifiées. Les modalités sont considérées comme *de re*, ce qui rend la théorie sujette aux mêmes critiques qui ont été adressées à Buridan par Guillaume d'Ockham. De plus, Al Fārābi défend à la suite d'Aristote deux sens du possible, mais ne réussit pas à séparer complètement les deux concepts de possibilité dans la mesure où la négation du possible bilatéral n'est pas donnée et est assimilée à l'impossibilité, ce qui crée, comme chez Aristote, de l'asymétrie et même une certaine incohérence. On peut également constater des divergences entre le premier traité et le deuxième, quoique ces divergences n'excluent pas une certaine continuité qui se manifeste à travers la reprise des mêmes définitions, ce qui tend à montrer que le deuxième traité est un approfondissement du premier et que les deux traités demeurent dans le sillage d'Aristote. Malgré ces insuffisances, la théorie est très riche, notamment au niveau des précisions qui sont apportées dans la formulation des oppositions modales et on peut légitimement considérer que ces précisions constituent un apport réel préfigurant celui des médiévaux.

Remerciements

Je remercie les éditeurs de cet ouvrage Amirouche Moktefi, Fabien Schang et Alessio Moretti pour m'avoir invitée à y présenter une contribution. Je suis aussi reconnaissante en particulier à Amirouche Moktefi et Fabien Schang pour leur aide précieuse et leurs conseils utiles ainsi qu'au professeur Stephen Read pour les articles qu'il a eu l'amabilité de m'envoyer.

Bibliographie

[1] Al Fārābi, A. N., 1988a, *Sharh Al 'Ibāra*, in *Al Mantiqiyāt li-al Fārābi*, vol. 2, textes réunis par Mohamed Teki Dench Proh, Edition Qom, 1409 de l'Hégire

[2] Al Fārābi, A. N., 1988b, *Al Qawl fi Al 'Ibāra*, in *Al Mantiqiyāt li-al Fārābi*, vol. 1, textes réunis par Mohamed Teki Dench Proh, Edition Qom, 1409 de l'Hégire

[3] Al Fārābi, A. N., 1988c, *Kitāb al Qiyās*, in *Al Mantiqiyāt li-al Fārābi*, vol. 1, textes réunis par Mohamed Teki Dench Proh, Edition Qom, 1409 de l'Hégire

[4] Aristote, 1969a, *De l'interprétation*, traduction française Tricot, Paris: Vrin

[5] Aristote, 1969b, *Les Premiers Analytiques*, traduction française Tricot, Paris: Vrin

[6] Avicenne, 1970, *Al Shifā*, volume 1, *Al Madkhal*, I. Madhkour (éd.), Le Caire

[7] Blanché, R., 1953, "Sur l'opposition des concepts", *Theoria*, vol. 19, p. 89-130

[8] Blanché, R., 1969, *Structures intellectuelles. Essai sur l'organisation systématique des concepts*, Paris: Vrin

[9] Blanché, R., 1970, *La logique et son histoire (d'Aristote à Russell)*, Paris: Armand Colin

[10] Chatti, S., 2012, "Logical Oppositions in Arabic Logic, Avicenna and Averroes", in J.-Y. Béziau & D. Jacquette (éd.) *Around and Beyond the Square of Opposition*, Birkhauser

[11] Czeżowski, T., 1955, "On Certain Peculiarities of Singular Propositions", *Mind*, vol. 64, p. 287-308

[12] Garson, J., 2009, "Modal Logic", *Stanford Encyclopedia of Philosophy*, E. N. Zalta (éd.)
http://plato.stanford.edu/entries/logic-modal/

[13] Knuutila, S., 2008, "Medieval Theories of Modalities", *Stanford Encyclopedia of Philosophy*, E. N. Zalta (éd.)
http://plato.stanford.edu/entries/modality-medieval/

[14] Lameer, J., 1994, *Al Fārābi and Aristotelian Syllogistics; Greek Theory and Islamic Practice*, Edition Brill

[15] Moretti, A., 2009, *The Geometry of Logical Opposition*, Thèse de Doctorat, Université de Neuchâtel

[16] Read, S., 2012, "The medieval Theory of Consequence", *Synthese*, 187, 899-912.

[17] Read, S. & Hodges, W., 2010, "Western Logic", *Journal of the Indian Council for Philosophical Research*, vol. 27, 1, p. 13-45

[18] Schang, F., 2010, "A structural Semantics for Multimodalities", présenté au 3e colloque N.O.T, Nice 22-23 Juin 2010

[19] Schang, F., 2011, "MacColl's Modes of Modalities", *Philosophia Scientiae*, vol. 15, Cahier 1, p. 150-188

Saloua CHATTI
Faculté des Sciences Humaines et Sociales de Tunis,
Université de Tunis, Tunisie

Formalism.
The Success(es) of a Failure
LIESBETH DE MOL[1]

1 Introduction

I had my first encounter with logic when I was still in high-school. My Dutch teacher chose to skip certain chapters in the course book including the one on logic. Fascinated as I was by things not taught in school I had a look at the chapter. Today I know that at that point I had no clue whatsoever what logic is. The chapter gave something in the spirit of the following syllogism as an example, adding that this was a correct derivation:

> All elephants are ruby red
> Bambi is an elephant
> _____
> Bambi is ruby red

I simply could not get this: how could this conclusion be correct? As is evident, the mistake I made was that I was focusing on the meaning of what is said rather than on the form of the deduction. I did not understand at that point in time that one can speak about truth in relation to form without reference to what *is* true, *empirically* speaking. Since that first encounter with logic, I have long disliked, if not opposed, the idea of formalization.

Nowadays I feel no longer appalled by logic-as-formalism, on the contrary. My interest however is *not* so much in the formalization of something "empirical" like human reasoning but in the form *as* form, form without content. This paper is an account of why "form" matters (to me) and why the quest for "meaningfulness" sometimes obscures and even slows down certain developments or ideas.

In the introduction to *Proofs and refutations* [Lakatos, 1976], an influential book in the philosophy of mathematics and science in general, Lakatos clarifies the motivation behind his book: to criticize and ultimately reject formalism as the *"latest link in the long chain of* dogmatist *philosophies*

---
[1]This research was supported by the Fund for Scientific Research, Flanders, Belgium and CNRS, UMR 8163 Savoirs, Textes, Langage, Université de Lille 3, France.

of mathematics" [Lakatos, 1976, p. 4]. By identifying formalism as the "bulwark" of logical positivism, he concludes that formalism somehow excludes the informal aspects of mathematics and that it denies mathematics its history. In this paper I will apply a Lakatos-inspired method on formalism itself viz. study *formalism-as* and embedded *in-a-practice* of (informal) mathematics. By doing so, I will argue that formalism as practiced is far removed from the kind of picture one gets from formalism upon reading Lakatos and that his view on formalism should at least be nuanced if one turns to actual formalism in action.

By way of a historically-inspired study of the work of a well-known mathematical logician, Emil Leon Post, I will make a stand for the concreteness, historicity and practicability of "form." It is shown that at least in the case of Post the formalist approach was a necessary prerequisite to unveil its fundamental limitations. Ultimately, this paper aims at showing how Post's formalism is relevant even today. By putting it into the perspective of computer science, it is suggested that it offers an interesting philosophical and practical alternative to the "art of simulation" as a means to explore the limits and possibilities of computation.

2 In search of the ultimate form

2.1 Lewis' influence on Post's early work

Perhaps one of the most extreme formalist convictions can be found in Chapter 6 *Symbolic Logic, Logistic and mathematical method* of Lewis' *Survey of Symbolic Logic*, a chapter that was removed from the later editions of [Lewis, 1918, pp. 355-56]:

> A mathematical system is any set of strings of recognizable marks in which some of the strings are taken initially and the remainder derived from these operations performed according to rules which are independent of any meaning assigned to the marks. [The] distinctive feature of this definiton lies in the fact that it regards mathematics as dealing, not with certain denoted things – numbers, triangles, etc. – nor with certain symbolized "concepts" or "meanings", but solely with recognizable marks, and dealing with them in such wise that it is wholly independent of any question as to what the marks represent. This might be called the "external view of mathematics" or "mathematics without meaning." [W]hatever the mathematician *has in his mind* when he develops a system, what he *does* is to set down certain marks and proceed to manipulate them [...]

With this lengthy quote, the stage is set for "pure" form, viz. form without meaning. As is clear, for Lewis, such form is in fact the ideal of mathematics *as an activity*.[1] Going a step further, if mathematics is the foundation of all of science then *"Logistik is the universal method for presenting exact science in ideographic symbols. It is the "universal mathematics" of Leibniz"* [Lewis, 1918, p. 372].

Imagine the impression this must have left on the young mathematician Emil Post. It is important to keep in mind that at that point in time, mathematical logic as a discipline hardly existed in the United States [Davis, 1995]. In fact, Emil Post would become one of the few U.S. mathematical logicians who, besides Alonzo Church, made fundamental contributions to the field in the 20s. Lewis' book was one of the few available English textbooks in circulation at that time, so it is not surprising that Post studied it.

In the same year as Lewis' book was published, Post was a postgraduate student at Columbia University. It was during that time that he was familiarized with the formal austerity of *Principia Mathematica* by Russel and Whitehead [Russell and Whitehead, 1913] through the teachings of Cassius J. Keyser. Together with Lewis' [Lewis, 1918] this would become the main influence on Post's Ph.D. *Introduction to a General Theory of Elementary Propositions* [Post, 1921].[2] Post however was not fully satisfied with the formal apparatus of *Principia* because [Post, 1921, pp.163–164]:

> [...] owing to the particular purpose the authors had in view they decided not to burden their work with more than was absolutely necessary for its achievements, and so gave up the generality of outlook which characterized symbolic logic. [W]e might take cognizance of the fact that the system of 'Principia' is but one particular development of the theory [and] so [one] might construct a general theory of such developments.

Hence, instead of working with *Principia* Post decided to develop his own formal apparatus, one, as Post would later write, that *"eschews all interpretation"* [Post, 1965].

---

[1] In the long footnote 17, p. 360, Lewis explains that this does not exclude creativity. The mathematician as the "manipulator" of the marks needs to be intelligent and ingenious for the derivation of required, interesting or valuable results. Lewis makes the analogy here with Gulliver *"who found the people of Brobdingnag (?) feeding letters into a machine and waiting for it to turn out a masterpiece. Well, masterpieces are combinations achieved by placing letters in a certain order! However mechanical the single operation, it will take a mathematician to produce masterpieces of mathematics."* (It is not in Brobdingnag but in the Academy of Lagado – where useless projects are undertaken – that Gulliver saw the machine for producing sentences and books.)

[2] See [Davis, 1994, De Mol, 2006, Urquhart, 2008].

2.2 The method of combinatory iteration

But why exactly did Post regard *Principia* as being too particular and why did he develop his own formal apparatus? The motivation behind this is what one could call (a kind of) methodological formalism: the development of the most general form of symbolic logic and ultimately mathematics as "*instruments of generalization*" which make possible a study of the *general* properties of the whole of mathematics. Post's idea was that if one wants to study the general properties of logic and mathematics then one needs not one particular system of symbolic logic or mathematics, but a general form that comprises all such possible systems. In this sense, Post's formalism can be regarded as a *method* to study (the foundations of) mathematics.

In an unpublished note from the *Emil Post Papers* held at the *American Philosophical Society* titled *Note on a Fundamental Problem in Postulate Theory* and dated June 4, 1921 Post makes explicit that formalism can be used "*to obtain theorems about all [possible] assertions*" of mathematics but that such a "*complete specification of the logic that is employed [in a mathematical system] is not made in the usual mathematical developments, and indeed is not necessary.*" In other words, Post did not aim nor expect to effectively replace the usual style of mathematics by formal logic; it was not his ambition to cleanse or cure mathematics from non-rigorousness or to get rid of its informality.

Even though Post's formalism can be called methodological, it is rooted in the belief that, ultimately, mathematics can be captured by form. This indeed does not necessarily mean that he expected that real-life mathematics would be replaced by formal proofs. What he did expect was that mathematics can be captured by (a) *general* form and that (b) by studying (particular instances of that) form it would become possible to prove theorems that say something about the whole of mathematics, about mathematics *in general*. An example of such a problem, which would in fact become his main focus in the period 1920-21 when he was a Procter fellow in Princeton, was what Post called *the finiteness problem* for first-order logic, viz. the famous Entscheidungsproblem proven undecidable by Church and Turing in 1936.

In its concrete realization, Post's formalism is completely in the spirit of Lewis' formalist philosophy. Indeed, in the same unpublished note just mentioned, Post identifies this "method" as the method of combinatory iteration and describes it as follows:

> [T]he method of combinatory iteration completely neglects [...] meaning, and considers the entire system purely from the symbolic standpoint as one in which both the enunciations and assertions are groups of symbols or symbol-complexes [....] and

where these symbol assertions are obtained by starting with certain initial assertions and repeatedly applying certain rules for obtaining new symbol-assertions from old.

How far this method of combinatory iteration would lead Post becomes clear if one connects Post's Ph.D. to his research during the period 1920-21 when he was a Procter fellow.

In his Ph.D. Post made a start with his method of generalization: he introduced the truth-table method for propositional logic (isolated from *Principia*) and proved that this logic is complete and consistent with respect to this method. He also emphasized that the truth table method provides a method that allows to decide the decision problem for the propositional calculus and generalized the two-valued truth table method to an arbitrary finite number of truth values hence laying the foundations for multi-valued logic. He also introduced a form intended as a *general* framework to reason about all systems of symbolic logic and, ultimately, mathematics. He referred to this form as *generalization by postulation* [Post, 1921, p. 176] and later called it the *canonical form A* [Post, 1943, Post, 1965]. It is a form that captures an infinite number of formal systems understood as finitary symbol manipulation systems. It resulted from a generalization of Post's formulation of propositional logic based on that from *Principia Mathematica*. Table 1 compares Post's formulation of propositional logic, using only the two logical functions $\sim, \vee$ with the canonical form A.[3,4]

Table 1: Comparison between Post's formulation of propositional logic and his canonical form A.

	Propositional Logic	**Canonical form A**
I.	If p is an elementary proposition then so is $\sim p$	If $p_1, \ldots, p_{m_1}$ are elementary propositions then so is $f_1(p_1, \ldots, p_{m_1})$
		$\vdots$
	If p and q are elementary propositions then so is $p \vee q$	If $p_1, \ldots, p_{m_\mu}$ are elementary propositions then so is $f_\mu(p_1, \ldots, p_{m_\mu})$
	Continued on next page	

---

[3] Note that I have not used the dots as brackets notation of Principia.

[4] Note that the description of propositional calculus in *Principia* is almost identical. However, it uses the three logical functions $\sim, \vee, \supset$. Remember that $p \supset q$ can be defined as $\sim p \vee q$ in propositional logic (See [Russell and Whitehead, 1913, p. 12]).

Table 1 – continued from previous page		
	Propositional Logic	**Canonical form A**
II.	The assertion of a function involving a variable p produces the assertion of any function found from the given one by substituting for p any other variable q, or $\sim q$, or $(q \vee r)$[5]	The assertion of a function involving a variable p produces the assertion of any function found from the given one by substituting for p any other variable q_i, or $f_1(q_1, \ldots, q_{m_1})$, ..., or $f_\mu(q_1, \ldots, q_{m_\mu})$
III.	$\vdash P$ $\vdash \sim P \vee Q$ produce $\vdash Q$	$\vdash \quad g_{11}(P_1, \ldots, P_{k_1}) \ldots \quad \vdash g_{k_1}(P_1, \ldots, P_{k_k})$ $\vdots$ $\vdash \quad g_{1k_1}(P_1, \ldots, P_{k_1}) \ldots \quad \vdash g_{kk_k}(P_1, \ldots, P_{k_k})$ produce produce $\vdash \quad g_1(P_1, \ldots, P_{k_1}) \ldots \quad \vdash g_k(P_1, \ldots, P_{k_k})$
IV.	Postulates: $\vdash \sim (p \vee p) \vee p$ $\vdash \sim (p \vee (q \vee r)) \vee (q \vee (p \vee r))$ $\vdash \sim q \vee (p \vee q)$ $\vdash \sim (\sim q \vee r) \vee (\sim (p \vee q) \vee (p \vee r))$ $\vdash \sim (p \vee q) \vee (q \vee p)$	Postulates: $\vdash h_1(p_1, p_2, \ldots, p_{l_1})$ $\vdash h_2(p_1, p_2, \ldots, p_{l_2})$ ... ... $\vdash h_\lambda(p_1, p_2, \ldots, p_{l_\lambda})$

As is clear from Table 1 what Post did was to extract some of the essential formal features of the postulational formulation of propositional logic and generalize them. Instead of the two logical functions $\vee, \sim$, a system in form A can have an arbitrary but finite number of functions; instead of having one production rule it can have an arbitrary number of production rules and instead of five postulates it can have an arbitrary but finite number of them. Post's formulation of propositional logic clearly fits canonical form A: it is just one of an infinite number of symbol manipulation systems that can be expressed in form A. The result is that Post now has a means to study not one but an infinite number of formal devices and hence study properties of mathematical systems as symbol manipulation systems *in general*.

Shortly after finishing his Ph.D., Post became a Procter fellow in Princeton: it was during that time that Post developed and studied several other forms, originally with the aim of proving that there is a *general* method to decide for any formula in first-order logic and ultimately *Principia* whether

---

[5]This corresponds to substitution.

or not it is derivable in that system, i.e., to prove the decidability of the finiteness problem for first-order logic, and, ultimately, for the whole of *Principia*. This was a very ambitious project. Indeed: *"Since Principia was intended to formalize all of existing mathematics Post was proposing no less than to find a single algorithm for all of mathematics"* [Davis, 1994]. Following the method of his Ph.D., his approach was to generalize and study meaningless form. In the introduction to *Account of an anticipation. Absolutely unsolvable problems and relatively undecidable propositions* [Davis, 1965], a manuscript which gives detailed descriptions on Post's research during 1920-21 and which was posthumously published in 1965 by Martin Davis,[6] Post explains the significance of this approach [Post, 1965, pp. 341–342]:

> Perhaps the chief difference in method between the present development and its more complete successors is its preoccupation with the outward forms of symbolic expressions, and possible operations thereon, rather than with logical concepts as clothed in, or reflected by, correspondingly particularized symbolic expressions, and operations thereon. [This] allows greater freedom of method and technique.

Instead of starting from the logic of *Principia*, Post decided to concentrate on his canonical form A, convinced that if one can work with a generalized form, stripped of meaning, it might be more easy or straightforward to prove the decidability of decision problems. Post knew that if he was able to prove the decidability of the finiteness problem for systems in canonical form A and if he would also be able to prove that first-order logic reduces to a system in canonical form A he would have succeeded in his ambitious goal of proving that any mathematical problem can be decided in a finite number of steps. Post was indeed able to prove that first-order logic as described in *Principia* can be reduced to a system in canonical form A (by way of a second form, canonical form B).[7] All that remained to be done now was to demonstrate the decidability of the finiteness problem for systems in canonical form A. Post's approach here was to start from the simplest (classes of) cases, by studying systems in *"which the primitive functions are all functions of one variable, the resulting relative simplicity of the systems allowing a direct analysis of the formal processes involved"* [Post, 1965, 346]. However, *"considerable further labor produced but minor dents in the problem for [systems in canonical form A] not so restricted."*

---

[6]This manuscript was offered for publication to *American Journal for Mathematics* but rejected by Hermann Weyl. A significantly abbreviated version of it was finally published as Post's influential [Post, 1943].

[7]See pp. 350–361 of [Post, 1965] for the details of the proof.

2.3 The frustrating problem of "tag"

So what to do next? Here things become a bit unclear, but it is known from [Post, 1943, Post, 1965] that the next *important* step forward in the method of combinatory iteration are Post's tag systems [De Mol, 2006].

Definition 1. (*v*-**tag system**) A tag system T consists of a finite alphabet Σ of μ symbols, a deletion number $v \in \mathbb{N}$ and a finite set of μ words $w_0, w_1, \ldots, w_{\mu-1} \in \Sigma^*$ called the appendants, where any appendant w_i corresponds to $a_i \in \Sigma$. A v-tag system has a deletion number v.

In a computation step of a tag system T on a word $A \in \Sigma^*$, T appends the appendant associated with the leftmost letter of A at the end of A, and deletes the first v symbols of A. This computational process is iterated until the tag system produces the empty word ϵ and hence halts. To give an example, let us consider the one tag system mentioned by Post [Post, 1943, Post, 1965] with $v = 3$, $0 \to 00$, $1 \to 1101$ [Post, 1943, Post, 1965]. If the initial word $A_0 = 110111010000$ we get the following productions:

~~110~~111010000
  ⊢ ~~111~~0100001101
    ⊢ ~~010~~00011011101
      ⊢ ~~000~~1101110100
        ⊢ ~~110~~111010000

The word A_0 is reproduced after 4 computation steps and is an example of a periodic word.

As explained by Post in his [Post, 1965], he arrived at tag systems when working on a problem related to but different from the finiteness problem and is now known as the unification problem [Davis, 1994]. This is the problem to determine for any two (logical) expressions what substitutions would make those two expressions identical. Post also found that tag systems and their decision problems are relevant for the finiteness problem of the canonical form A. Hence, "*[tag systems] appeared as a vital stepping stone in any further progress to be made*" [Post, 1965, p. 361].

If we compare the formal definition of tag systems with that of systems in canonical form A, it is clear that whereas the canonical form still bears a clear relation with propositional logic, this is not the case for tag systems. It was Post's hope that his study of tag systems would in fact be a step towards a solution for the finiteness problem for systems in canonical form A. More specifically, Post hoped to tackle what he called the problem of "tag" for which he formulated two variants nowadays known as the halting and reachability problem for tag systems [De Mol, 2010]:

Definition 2. The halting problem for tag systems is the problem to determine for a given tag system T and any initial word A_0 whether or not T will halt when started from A_0.

Definition 3. The reachability problem for tag systems is the problem to determine for a given tag system T, a fixed initial word A_0 and any arbitrary word $A \in \Sigma^*$, whether or not T will ever produce A when started from A_0.

Post would spend nine months of research on these devices. His approach was to start from the simplest cases, and, if successful, try to "scale" the results for the simplest cases to the complete class of tag systems. He was able to prove that the class of tag systems with $v = \mu = 2$ has a decidable halting and reachability problem, a proof which *"involved considerable labor"*. He considered it as the *"major success"* of his Procter fellowship.[8] However, for cases that seemed to go but one step beyond the case $v = \mu = 2$, Post only found *"intractable"* cases and cases of a *"bewildering complexity"* [Post, 1965, 382]. In the end, even though Post initially had been quite optimistic about the possibility of successfully proving the problem of "tag" decidable, it was his meeting and interaction with this form that ultimately led to the reversal of his entire program of proving the finiteness problem decidable [Post, 1965, p. 363]:

> For a while the case $v = 2$, $\mu > 2$, seemed to be more promising, since it seemed to offer a greater chance of a finely graded series of problems. But when this possibility was explored in the early summer of 1921, it rather led to an overwhelming confusion of classes of cases, with the solution of the corresponding problem depending more and more on problems in ordinary number theory. Since it had been our hope that the known difficulties of number theory would, as it were, be dissolved in the particularities of this more primitive form of mathematics, the solution of the general problem of "tag" appeared hopeless, and with it our entire program of the solution of finiteness problems. This *frustration* [my emphasis], however, was largely based on the assumption that "tag" was but a minor, if essential, stepping stone in this wider program.

Post had clearly underestimated the complexity a simple form such as that of "tag" can 'generate'. Instead of being convinced of the existence of an ultimate method to decide all of mathematics, he now considered the possibility that this might be a hopeless ambition since already this "primitive form of mathematics" results in major difficulties.

---

[8] And I regard it as one of my own major successes to have reproven this result (See [De Mol, 2010]). Note that (developing) this proof also involved considerable labor.

2.4 The actual reversal of Post's programme

After his frustrating experience with tag systems, Post developed two other important forms during his Procter fellowship: systems in *canonical form* C, which are nowadays known as Post production systems in the context of formal language theory, and the normal form. I only describe the normal form here.

Systems in normal form, normal systems for short, are a special class of systems in canonical form C. A system in normal form has only one initial word (postulate) and a finite set of production rules all of the following form:

$$g_i P_i$$
produces
$$P_i g'_i$$

Clearly, normal systems are very similar to tag systems. In fact, the production rules of a tag system are easily rewritten in normal form.

Knowing from his experience with tag systems that apparent formal simplicity does not necessarily imply real simplicity, Post started on a project of proving the "power" of systems in normal form, viz. their generality: he first proved that canonical form A and B can be reduced to a system in canonical form C and then, most importantly, proved that the canonical form C reduces to normal form. This fundamental result was later published as [Post, 1943]. From this Post concluded that in fact the whole of *Principia* and hence mathematics could perhaps be reduced to the normal form:

> [F]or if the meager formal apparatus of our final normal systems can wipe out all of the additional vastly greater complexities of canonical form B, the more complicated machinery of the latter should clearly be able to handle formulations correspondingly more complicated than itself.

This insight resulted in the formulation of what Martin Davis has called Post's thesis:

Post's thesis *Every generated set of sequences on a given set of letters $a_1, a_2, ..., a_\mu$ is a subset of the set of assertions of a system in normal form with primitive letters $a_1, a_2, ..., a_\mu, a'_1, a'_2, ..., a'_\nu$, i.e., the subset consisting of those assertions of the normal system involving the letters $a_1, a_2, ..., a_\mu$.*

Post's thesis identifies the vague notion of generated (set of) sequence(s) with generated by a normal system. Even though this thesis is quite technical in nature, it is logically equivalent to Turing's more famous thesis.

Post soon understood the implications of this thesis. He had already learned from tag systems that his program of proving the whole of mathematics decidable might in fact be hopeless. He was now able to prove with the diagonal method that there is no finite method to decide for any normal system and some word whether or not that word can be generated by that normal system. Since he was convinced of the generality of normal systems he concluded that there are *absolutely* unsolvable problems. Post even went one fundamental step further and derived, on the basis of these results, that no logic is complete hence anticipating part of Gödel's results be it without formal proofs.

2.5 "I study mathematics as a product of the human mind"

Having established a thesis logically equivalent to Church's and Turing's 15 years before the facts, Post understood that even though he was now convinced of the universality of normal form [Post, 1965, p. 387]:

> [for the thesis to obtain its full generality] an analysis should be made of all the possible ways the human mind can set up finite processes to generate sequences.

This view is very similar to what Turing would later write in his famous 1936 paper *On computable numbers* [Turing, 1936] where he states that:

> The real question at issue is: "What are the possible processes which can be carried out in computing a number?"

Although hardly ever acknowledged in the literature, in 1921 Post was quite aware of the significance of what Turing calls the "real question at issue." In fact it can be argued that Post made a start with such an analysis as early as 1921-22, an analysis which lay at the basis of Post's note from 1936 [Post, 1936] and contains a formulation which is almost identical to Turing machines. How else does one explain that both Post and Turing developed quasi-identical formalisms?[9]

---

[9]The main reason why Turing's thesis is considered in the literature as superior is exactly because of this analysis of the processes involved when the mathematician is computing a number, an analysis which, by eliminating all the non-essential features of this process, resulted in the well-known Turing machine. Even though I value Turing's work very highly, it is my view that several recent historical and philosophical studies on the topic are too much biased in their high praise for Turing's work against Church's and especially Post's. I have found no satisfying reason in the literature to regard Turing's thesis as being superior to Post's second thesis (see below). It is not because Post's

This 1936 note was submitted by Post to the Journal of Symbolic Logic after having read Church's 1936 paper. It contained a thesis which identifies the vague notion of *solvability* of a problem with solvability by his *formulation 1*. Although almost identical to Turing machines there is one important and philosophical difference between Post's and Turing's approaches: Turing's analysis is one of the mathematician in the process of computing a number, for Post it is an analysis of the possible *mental processes* involved when generating a set and, later in 1936, when solving a decision problem. This is reflected in the fact that Turing's formalism is in terms of idealized computing machines, whereas Post's was in terms of sets of instructions in a formal language (see [Davis, 1994]). The fact that for Post his theses are related to human mental processes is reemphasized in his note [Post, 1936, p. 105]:

> Its purpose [of formulation 1] is not only to present a system of
> a certain logical potency but also, [...] of psychological fidelity

It is exactly for this reason that Post could not agree with Church on the idea of regarding his thesis (or any other logically equivalent one) as being but a formal definition of a vague, intuitive concept.[10] It is also why for Post his thesis should be understood as a *working hypothesis* and, in case more and more support could be found for it, a *natural law*. Indeed, for Post his thesis is about the human mind and its mathematical capabilities, hence [Post, 1936, p. 103]:

> [...] to mask this identification under a definition hides the fact
> that a fundamental discovery in the limitations of the mathe-
> maticizing power of *Homo Sapiens* has been made and blinds us
> to the need of its continual verification.

This is a very strong philosophical point of view not only with respect to the thesis but also with respect to mathematics in general. It shows that even though Post can be considered as a formalist, this does not mean that he understood mathematics and its formalizations as something that can be isolated from humans, a point made even more explicit here [Post, 1965, p. 403]:

> I study mathematics as a product of the human mind not as
> absolute.

---

paper does not contain the (philosophical) analysis nor the major results of Turing's [Turing, 1936] that the thesis as such would be worth less.

[10] For instance, in Church's thesis, effective calculability is defined as λ-definability and general recursive functions.

Does the conclusion of a fundamental limitation of the "method of combinatory iteration" mean that Post had turned his back to symbolic logic and "form"? No. On the contrary, the normal form would remain a fundamental form throughout his work. He even used it as the formal framework in his founding paper on recursive functions [Post, 1944]. More important here, given the discovered limitations, it is symbolic logic itself that can be used as a method to explore and develop these limitations [Post, 1965]:

> [...] the creativeness of human mathematics has a counterpart inescapable limitation thereof – witness the absolutely unsolvable (combinatory) problems. Indeed, with the bubble of symbolic logic as universal logical machine finally burst, a new future dawns for it as the indispensable means for revealing and developing those limitations. For [...] Symbolic Logic may be said to be Mathematics become self-conscious.

In a letter to Church dated March 24, 1936 a similar point is made:[11]

> For if symbolic logic has failed to give wings to mathematicians this study of symbolic logic opens up a new field concerned with the fundamental limitations of mathematics, more precisely the mathematics of Homo Sapiens.

To be clear: this particular view does not imply that Post somehow supported computationalism, viz., the idea that the mind is like a Turing machine.[12] It only means that there are things we cannot do (at least if we indeed interpret the thesis as something that relates to human activity) and that it is symbolic logic than can be used to study the boundaries of these human limitations.[13]

3 Re: some high-speed logic. A discussion

The fact that Post emphasized that his thesis (and those that are logically equivalent to it) is a hypothesis because, if true, it implies a discovery of

---

[11] The letters from Post to Church can be found in the Alonzo Church papers, box 20, Folder 14; Department of Rare Books and Special Collections, Princeton University Library.

[12] It should be noted here that after his discoveries he became more and more convinced of the significance of mathematical creativity. On several occasions he pleaded for a mathematics that is more informal (!) and less axiomatic. In fact in the introduction of [Post, 1965, p. 343] he even makes a plea for *a reversal of the entire axiomatic trend [...] with a return to meaning and truth.*"

[13] Regretfully this is not often not what the debate on the Church-Turing thesis focuses on nowadays. If one takes some of the statements by people like Copeland seriously one gets the impression that they want to deny us these very limitations.

a fundamental human limitation, is most probably rooted in Post's explorations of "form" in the period 1920-21 and the reversal of his program that resulted from it. Indeed, quite unlike Turing who started out from the idea of formalizing the vague notion of computability, Post formulated his thesis on the basis of a profound analysis of systems of symbolic logic.[14] The insight that something as simple as tag systems cannot be controlled confronted Post with the limits of finite methods and, since these methods are human, also with his own limitations. Hence, Post's formalist approach ultimately resulted in a view on symbolic logic that seems far removed from the kind of picture one gets from Lakatos' reading of formalism. Here is an excerpt from his [Lakatos, 1976]:

> But what can one *discover* in a formalized theory? [...] *First*, one can discover the solution to problems which a suitably programmed Turing machine could solve in a finite time [...]No mathematician is interested in following out the dreary mechanical 'method' prescribed by such decision procedures. *Secondly*, one can discover the solutions to problems (such as: is a certain formula in a non-decidable theory a theorem or not?), where one can be guided only by the 'method' of unregimented insight and good fortune. Now this bleak alternative between the rationalism of a machine and the irrationalism of blind guessing does not hold for live mathematics

In the light of Post's struggle with "form" which resulted in a philosophical point of view that understands symbolic logic as the means to develop and explore the limits of mathematics and its formalizations, but also as *"mathematics become self-conscious"*, this black-and-white picture of formalism vs. informal mathematics should at least be nuanced.[15] Furthermore, even at the time when Post was still a full-blood formalist he did not expect that *real-world* mathematics would be replaced by formalism (See Sec. 2.2, p. 32).

If there is one thing one can learn from Post's formalism, it is that it is formalism *itself* and *in practice* that makes possible the study of its very own

---

[14]This aspect of Post's early work is quite parallel to the way Church arrived at the first formulation of his thesis in which he identifies calculability with λ-definability. It was only by studying (properties of) λ-calculus and understanding its power that Church first came to the idea of defining (in his view) the vague notion of calculability.

[15]Even though I adopt the philosophical view on mathematics which is historically-embedded and practice-based – a view which has been influenced by Lakatos' [Lakatos, 1976], the way formalism is described by Lakatos is very much a-historical and at best caricatural. The fact that Tarski, Curry, Church et al. are put on the same line as the logical positivists seems unfair. For instance Curry's formalism is much more delicate than the image one gets from Lakatos' Curry [Curry, 1951].

limitations. In this sense, Lakatos' question *But what can one discover in a formalized theory?* gets a very different answer from the one provided by Lakatos. It is Post's formalist method of simplifying through generalizations that led to his results and philosophical point of view and it is hard to imagine that Post would have called this method one of "unregimented insight and good fortune", formalist though he was at that time.

Post's story shows us that it was exactly the Lewisean heterodox view on mathematics, a mathematics stripped of all meaning, that resulted in the anticipation of the fundamental results of the 30s by Gödel, Church and Turing, even though it was not published at the time. Indeed, whereas [Davis, 1982]:

> Hilbert and his school went on to approach the decision problem for quantification theory semantically, Post evidently felt that was not a promising direction because the combinatorial intricacies of predicate logic were too great to penetrate into that manner, and what he proposed instead was to simplify through generalization.

Some 10-15 years after Post, the formalist school of thought would officially achieve the height of its (own) failure. The very limitations already discovered by Post in 1921 were now proven in detail and published. Gödel's incompleteness results are often seen as the death knell of the Hilbertian optimism so famously (and ironically) captured in Hilbert's epitaph "wir müssen wissen, wir werden wissen". It showed that no finite axiomatic system would ever be able to capture the whole of mathematics. Some five years later it would be up to Church and Turing to furthermore prove that no finite (formal) method will ever be found which is able to decide problems logically equivalent to the Entscheidungsproblem of first-order logic.

Despite the failure of the formalist program in the sense of Hilbert, it is not the case that formalism was dead and buried after that. In fact, out of the ashes of the failure some of the foundations would be laid for a new discipline to be: computer science. Indeed, with the rise of the electronic and programmable computer it became clear that the formal devices developed by Church, Post, Turing et al. were in fact very useful. Hence, the results of that which is often regarded as an abstract and old-fashioned philosophy of mathematics, attained a new and vigorous life in the context of the machine that we all use on an everyday basis.

Is this somehow surprising? In a sense it is not. One should not forget that the computer can in fact be understood as the physical realization of "calculability" and is hence the physical pendant of the forms developed by Church, Post and Turing. In this sense, the computer can also be understood as a machine without meaning, at least to some extent. That the computer is a machine without meaning, a machine that does not really understand in the way we are able to understand, is in fact one of the classical arguments of those who are against the idea of an intelligent machine, and, quite often, consciously or unconsciously, in favor of a pejorative and derogatory view of the machine. On the other side of the spectrum there are those who are trying to understand how the machine can be made (more) intelligent and/or (more) natural mostly by focusing on simulation.

If we look at what is done with computers nowadays, the least one can say is that these machines are quite influential in our everyday and professional lives. Understood as machines without meaning who can only "understand" form but not meaning, this would mean that it is mechanized "formal logic" that we all so much depend on. Of course, when we are interacting with the machine, we are hardly aware of this. This is due to the fact that it is the explicit purpose of software developers to create an illusion of meaningfulness made possible by adding many layers on top of the bare electrical pulses of the machine so that the user does not need to be bothered with the technicalities of the machine, all, of course, for the sake of "user-friendliness". In the meantime philosophers keep debating for or against the 'art of simulation'.

Few, however, are taking up the challenge posed by Derrick H. Lehmer, a number theorist and computer pioneer, in his paper *Some high-speed logic* [Lehmer, 1963]: instead of trying to let the machine excel in the art of simulation, or criticize it because it is poor at mimicking us, we should perhaps start to take seriously the idea of having a fair contest/interaction, one in which the machine is allowed to do what it is good at. Taking such a challenge philosophically seriously, Post's formalism put into a modern perspective could be one possible approach. It was argued here that Post dismissed meaning convinced that by focusing instead on the formal aspects/structure of mathematics it would be possible to understand some fundamental properties of the whole of mathematics. It was this approach which allowed Post to take form seriously, to explore it and to uncover not only its possibilities but also its limitations. Similarly, it is perhaps by interacting with and studying the computer as a machine without meaning, stripped of its simulated "semantics", that we will be able to understand and explore the limitations and possibilities of computation in a context averse to the philosophically laden idea of the mimicking machine.

BIBLIOGRAPHY

[Curry, 1951] Curry, H. B. (1951). *Outlines of a Formalist Philosophy of Mathematics*. North Holland, Amsterdam.
[Davis, 1965] Davis, M. (1965). The Undecidable. Basic Papers on Undecidable Propo-

sitions, Unsolvable Problems and Computable Functions. Raven Press, New York. Corrected republication (2004), Dover publications, New York.

[Davis, 1982] Davis, M. (1982). Why Gödel didn't have Church's Thesis. *Information and Control*, 54:3–24.

[Davis, 1994] Davis, M. (1994). Emil L. Post. His life and work. in: [Davis, 1994b], xi–xviii.

[Davis, 1994b] Davis, M. (1994). *Solvability, provability, definability. The collected works of Emil L. Post*, Birkhauser.

[Davis, 1995] Davis, M. (1995). Logic in the twenties. *The Bulletin of Symbolic Logic*, 1(3):273–278.

[De Mol, 2006] De Mol, L. (2006). Closing the circle: An analysis of Emil Post's early work. *The Bulletin of Symbolic Logic*, 12(2):267–289.

[De Mol, 2010] De Mol, L. (2010). Solvability of the halting and reachability problem for binary 2-tag systems. *Fundamenta Informaticae*, 99(4):435–471.

[Lakatos, 1976] Lakatos, I. (1976). *Proofs and Refutations. The logic of mathematical discovery*. Cambridge University Press, Cambridge.

[Lehmer, 1963] Lehmer, D. H. (1963). Some high-speed logic. In *Experimental Arithmetic, High Speed Computing and Mathematics*, volume 15 of *Proceedings of Symposia in Applied Mathematics*, pages 141–376.

[Lewis, 1918] Lewis, C. I. (1918). *A survey of Symbolic Logic*. University of California Press, Berkeley.

[Post, 1921] Post, E. L. (1921). Introduction to a general theory of elementary propositions. *American Journal of Mathematics*, (43):163–185.

[Post, 1936] Post, E. L. (1936). Finite combinatory processes - Formulation 1. *The Journal of Symbolic Logic*, 1(3):103–105. Also published in [Davis, 1965], 289–291.

[Post, 1943] Post, E. L. (1943). Formal reductions of the general combinatorial decision problem. *American Journal of Mathematics*, 65(2):197–215.

[Post, 1944] Post, E. L. (1944). Recursively enumerable sets of positive integers and their decision problems. *Bulletin of The American Mathematical Society*, (50):284–316.

[Post, 1965] Post, E. L. (1965). Absolutely unsolvable problems and relatively undecidable propositions - Account of an anticipation. in: [Davis, 1995], 340–433. Also published in [Davis, 1994b].

[Robinson, 1965] Robinson, J. A. (1965). A machine-oriented logic based on the resolution principle. *Communications of the ACM*, 5:23–41.

[Russell and Whitehead, 1913] Russell, B. and Whitehead, A. N. ((1910, 1912, 1913)). *Principia Mathematica, vol. I-III.* Cambridge University Press, Cambridge. Unabridged, Digitally Enlarged Printing Of Volume I of III published by Merchant Books, 2009.

[Turing, 1936] Turing, A. M. (1936-37). On computable numbers with an application to the Entscheidungsproblem. *Proceedings of the London Mathematical Society*, (42):230–265. A correction to the paper was published in the same journal, vol. 43, 1937, 544–546. Both were published in [Davis, 1965], 116–151.

[Urquhart, 2008] Urquhart, A. (2008). Emil Post. In Gabbay, D. M. and Woods, J., editors, *Handbook of the History of Logic. Volume 5. Logic from Russel to Church.* Elsevier.

Liesbeth DE MOL
CNRS, UMR 8163 Savoirs, Textes, Langage
Université de Lille 3, France

La Quadrature du Carré
GEORGE ENGLEBRETSEN

1. Une logique du plus/moins

Nous commençons par la construction d'un langage artificiel très simple. Son vocabulaire se compose d'un stock illimité de termes élémentaires, chacun représenté par une lettre (d'ordinaire majuscule) de l'alphabet romain. Le langage est doté aussi d'une paire de "foncteurs de termes", expressions qui, lorsqu'elles sont appliquées à un terme ou une paire de termes, forment un nouveau terme composé. Ce foncteur est unaire, et il est représenté par le signe moins: $-$. L'autre foncteur est binaire, et il est représenté par le signe plus: $+$. Le foncteur unaire est placé devant un terme pour former un nouveau terme. Le foncteur binaire est placé entre une paire de termes pour former un nouveau terme. Par exemple, si A, B et C sont des termes alors ceux qui suivent le sont également (où les parenthèses servent de ponctuations conformes à l'usage commun): $-A$, $-B$, $A+C$, $(-A)+B$, $(-C)+(-B)$, $-(C+A)$, $(-B)+(A+B)$, $-((A+B)+(C+A))$. Les termes composés formés à partir du signe plus ont certaines propriétés formelles caractéristiques. Le foncteur binaire est une relation symétrique et associative entre des termes. $A+B$ et $B+A$ sont équivalents, par exemple; de même pour $(A+B)+C$ et $A+(B+C)$. Le plus n'est pas réflexif et n'est pas transitif.

Difficile d'imaginer un langage formel plus simple. C'est un langage qui n'est pas utile à un grand nombre de choses. Cela dit, il reflète bien une petite portion de notre langage naturel. Si nous nous en tenons à la langue française, nous y trouvons effectivement des formes d'expressions qui sont symétriques et associatives mais ne sont ni réflexives ni transitives. Nous disons: "Tom est riche et heureux", où le terme composé "riche et heureux" pourrait tout aussi bien être remplacé par "heureux et riche". Un terme tel que "intelligent et beau mais vaniteux" en dit autant que "intelligent mais beau et vaniteux". Les termes de conjonction partagent ici les mêmes propriétés formelles de notre plus. Nous disons aussi: "Tom est riche et Sarah est vaniteuse", ce que l'on pourrait tout aussi bien dire par "Sarah est vaniteuse et Tom est riche". En d'autres termes, nous conjoignons (d'habitude avec un mot comme "et") à la fois des paires de termes et des énoncés entiers (c'est-à-dire des formules propositionnelles). Et dans la plupart des cas de ce genre, la conjonction possède les propriétés formelles du plus de notre langage artificiel. Ceci laisse entendre que, pour aller vite, nous pourrions interpréter notre foncteur plus comme un signe de conjonction puisque les conjonctions partagent les propriétés formelles intégrées au plus. Il y a d'autres types d'expressions que nous utilisons en temps normal et qui ne sont pas des conjonctions, mais qui partagent les propriétés formelles du plus. Considérons l'énoncé "Quelques logiciens sont des artistes". Ici, une opération est effectuée sur les termes "logiciens" et "artistes" par l'expression "quelques ... sont ..." pour former l'énoncé. L'expression "quelques ... sont ..." est une expression qui s'applique à des paires de termes pour former des énoncés. C'est un foncteur de termes du langage naturel. De plus, il possède les propriétés formelles de notre plus.

Un mot sur notre moins, avant de poursuivre. Il est évident que les signe plus et moins de notre langage sont destinés à nous rappeler les mêmes signes tels qu'ils apparaissent en arithmétique ou en algèbre. L'addition des mathématiques élémentaires partage la totalité des propriétés formelles de notre plus. Notre moins est similaire également au moins mathématique que nous connaissons bien. Il peut être interprété à travers notre langage naturel comme étant la négation, et l'action de nier est une des choses que nous tous savons très bien faire. Nous nions des termes simples (ainsi utilisons-nous "non marié","non partisan", "pas intelligent", etc.), nous nions des termes composés ("ni riche ni pauvre"), nous nions des énoncés entiers ("Edouard n'est pas très malin", "Ce n'est pas une belle journée", "Il n'y avait pas âme qui vive").

Notez que ce qui a été dit jusqu'ici à propos des termes se dit également à propos des énoncés. De notre point de vue "terministe", en effet, un énoncé n'est autre qu'un terme composé (mais tous les termes composés ne sont pas des énoncés, bien sûr). Jusqu'à présent, nous disposons de ce petit langage formel qui peut être utilisé pour modéliser une petite partie de la logique de notre langage naturel. Ce n'est pas une grande partie de la logique, bien sûr, dans la mesure où elle repose sur les maigres propriétés formelles du plus. Il est donc temps d'enrichir notre langage formel, en lui donnant de nouveaux éléments qui nous permettront non seulement de modéliser davantage de types d'expressions du langage naturel mais aussi de modéliser nos manières typiques de manipuler notre langage dans le processus de raisonnement. Nous avons vu plus haut qu'un énoncé tel que "Quelques logiciens sont des artistes" conjoint une paire de termes par le biais de l'expression formative "quelques ... sont ...". Et nous avons vu "quelques ... sont ..." pouvait être formulée par notre plus binaire. La chose étrange (une des choses étranges, à dire vrai), ici, c'est que le plus est une expression singulière alors que "quelques ... sont ..." se compose d'une paire de mots. De plus, il serait normal de rechercher une façon similaire de formuler des énoncés formés par la conjonction de paires de termes au moyen d'une expression légèrement différente: "tous ... sont ...". Commençons donc l'enrichissement de notre langage formel en *scindant* le plus binaire en deux parties, l'une représentant des expressions telles que "quelques" et l'autre représentant des expressions telles que "sont". Notre plus binaire scindé se composera désormais de deux signes séparés (comparable aux "quelques" et "sont", dans "quelques ... sont ..."). Le binaire scindé sera: + ... + ... Nous ne sommes pas forcés d'abandonner le + non scindé. Des exemples de version française de notre binaire scindé sont "quelques ... sont ...", "un(e) ... est ...", " ... et ...". Il est important de garder soigneusement à l'esprit le fait que les deux parties de notre fonctor scindé ne constituent pas eux-mêmes des fonctors. Les termes composés (parmi lesquels figurent les énoncés) ne peuvent pas être formés d'une paire de termes syntaxiquement plus simples avec une seule partie d'un formatif scindé. Les deux parties sont nécessaires. Par simple souci de commodité, nous appellerons *quantificateur* la première partie de ce formatif binaire scindé et *qualificateur* la seconde partie. Si nous voulons construire un langage formel destiné à nous permettre de modéliser nos manières de raisonner par l'intermédiaire d'un langage naturel, nous devons admettre au moins un autre type de quantificateur.

Considérons un énoncé simple comme "Quelques logiciens sont des artistes", que l'on pourrait formuler de façon simple sous la forme "+L+A". Un énoncé étroitement lié nierait le second terme: "Quelques logiciens ne sont pas des artistes", qui serait formulé sous la forme "+L+(−A)". Puisque tout énoncé est un type de terme composé, nous pouvons nier les énoncés de la même façon que nous nions des termes. Les négations correspondantes de nos deux énoncés seraient "−(+L+A)" et "−(+L+(−A))". Ces deux formes peuvent être interprétées par "Ce qui suit n'est pas le cas: quelque logicien est un artiste" et "Ce qui suit n'est pas le cas: quelque logicien n'est pas un artiste", qui constituent des énoncés de la langue française inélégants d'un point de vue grammatical mais corrects. Mais ils peuvent être reformulés en "Aucun logicien n'est un artiste" et "Aucun logicien n'est pas un artiste". Le premier des deux est tout à fait naturel, le second l'est moins. Ils pourraient être paraphrasés encore en "Tous les logiciens ne sont pas des artistes" et "Tous les logiciens sont des artistes". Or le second semble plus naturel que le premier. Identifions tous les énoncés plus naturels parmi les quatre: "Quelques logiciens sont des artistes", "Quelques logiciens ne sont pas des artistes", "Aucun logicien n'est un artiste", et "Tous les logiciens sont des artistes". Le troisième est la négation du premier, tandis que le quatrième est la négation du second. Mais le troisième et le quatrième n'ont plus l'apparence de négations. Nous pouvons rectifier ceci en prenant les formules négatives de ces deux énoncés, "−(+L+A)" et "−(+L+(−A))", puis en introduisant (comme en algèbre) le signe moins initial dans la formule pour obtenir: "−L−A" et "−L−(−A)". Mais que devons-nous faire des signes formatifs, ici? Considérons "L−(−A)" à la lumière du fait que les termes composés (qui incluent les énoncés) sont formés de paires de termes au moyen d'un foncteur binaire. Nous scindons notre foncteur plus binaire pour obtenir + ... + ... (interprétable par exemple en "quelques ... sont ..."). Il est clair que les deux premiers signes moins de "−L−(−A)" constituent un foncteur binaire joignant les deux termes "L" et "−A" pour former l'énoncé. Mais que signifient-ils (quels sont leurs analogues dans le langage naturel)? Empruntons un peu plus à l'algèbre. Nous pouvons dire que (i) toute paire adjacente de signes similaires peut être remplacée par un signe plus, et que (ii) toute paire adjacente de signes opposés peut être remplacée par un moins tant qu'aucun des deux signes n'est un quantificateur (tant que le second est un foncteur unaire et que le premier est soit un foncteur unaire également, soit un qualificateur). Mais patience; jusqu'ici, notre seul quantificateur est le premier plus du foncteur binaire scindé + ... + ..., qui représente des mots comme "quelques". En identifiant "−L−(−A)" à "−(+L+(−A))", nous avons défini en effet un nouveau quantificateur (qui représente des mots comme "tout"). Le fait d'admettre l'élimination des second et troisième signe moins (ce qui donne un plus) nous donne la formule "−L+A", qui peut être interprétée comme "Tout logicien est un artiste", avec "− ... + ..." constituant un nouveau foncteur binaire scindé défini ("tout ... est ..."). Alors que la relation logique représentée par "+ ... + ..." est symétrique et associative, la nouvelle relation "− ... +" est réflexive et transitive.

2. Le carré des oppositions

Il y a bien plus à dire à propos de ce langage formel légèrement enrichi. Il s'avère qu'il peut être augmenté d'un certain nombre de manières de produire un langage formel réellement puissant et adéquat pour les besoins d'un langage formel. Il est puissant du point de vue de son expression (capable de formuler une très grande variété de types d'expressions du langage naturel), et il est puissant du point de vue de ses inférences (capable de modéliser un grand nombre de nos manières ordinaires de raisonner correctement). Mais tout cela a été dit et défendu à de nombreux endroits par un certain nombre de personnes. Ce que nous voulons faire, ici, c'est demander plutôt si cette manière de formuler un langage formel peut apporter un nouvel éclairage sur un thème logique très ancien. Le thème dont il s'agit est le Carré des Oppositions et les relations logiques entre ce qu'Aristote appelait "les quatre", qui sont censées être représentées dans le carré.

Le carré "traditionnel" est destiné à offrir une claire illustration visuelle des diverses relations logiques au sein de la famille d'énoncés qui suit (où les termes sont remplacés par des lettres de termes, qui procèdent comme des variables en vue de produire un schéma d'énoncé plutôt qu'un énoncé réel du langage naturel):

1. Quelques S sont P
2. Quelques S ne sont pas P
3. Aucun S n'est P
4. Tout S est P

Chacun de ces énoncés a reçu un nom spécifique. Les trois premiers ont été appelés (dans l'ordre) I, O, et E. Bien que les logiciens antiques aient appelé A le quatrième, nous l'appellerons a. Pas de panique; le nom antique "A" ne sera pas tout à fait perdu. Un carré traditionnel ressemblerait (presque) à ceci:

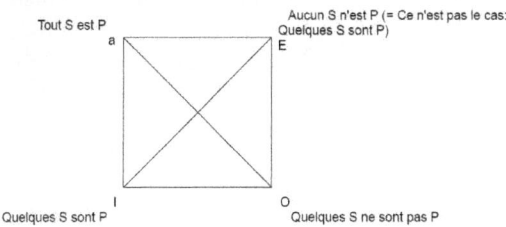

Remplaçons les énoncés par leurs formulations dans notre langage artificiel.

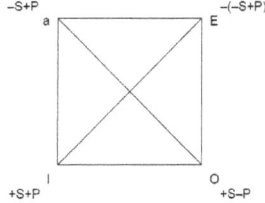

D'après la tradition, les formes situées aux angles diagonalement opposés du carré sont des *contradictoires* logiques. Cela veut dire que, pour une paire d'énoncés de ce genre, une au moins est vraie (la "loi du tiers exclu") et ce ne peut pas être le cas que les deux soient vrais (la "loi de non-contradiction"). Les deux formes du haut ont été qualifiées de *contraires* logiques (elles peuvent être fausses toutes les deux, mais elles ne peuvent pas être vraies toutes les deux). Les deux formes du bas ont été qualifiées de *subcontraires* (elles pourraient être vraies toutes les deux, mais elles ne pourraient pas être fausses toutes les deux). Pour finir, la tradition a soutenu généralement l'idée que chacune des formes du haut entraîne logiquement la forme située en dessous d'elle, ces dernières étant qualifiées de *subalternées* des premières.

Le carré traditionnel est une belle chose. Or comme nous le savons tous (et comme nous le disent aujourd'hui biologistes et psychologues), nous avons tendance à trouver davantage de beauté dans la symétrie que dans l'asymétrie. Un regard attentif sur le carré traditionnel révèle une once d'asymétrie. On pourrait s'attendre à ce que les contradictoires équivalent à des négations mutuelles (ou que l'une est la négation de l'autre, dans notre logique du plus/moins). Les formes I et E sont manifestement des contradictoires. Etant donné que "aucun" est une contraction de "ce n'est pas le cas que quelque/un(e)", E est la négation de I. Mais que dire de la paire contradictoire de O et *a*? Notez que, pour les formes I, O et E, le seul quantificateur qui apparaît est "quelque(s)" (+). La négation *proprement dite* de O serait "−(+S−P)" ("Ce n'est pas le cas que quelque S n'est pas P", "Aucun S n'est pas P"). Cette forme est celle qui mérite le nom A.

Un "Carré des Oppositions Fondamental" n'afficherait de formes propositionnelles que celles faisant usage du quantificateur "quelque(s)" (+). En d'autres termes:

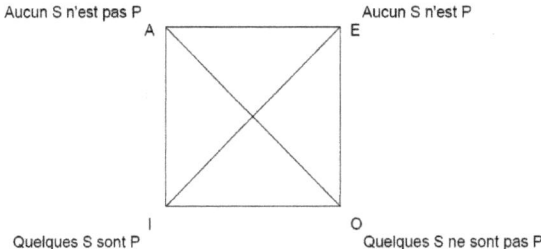

Ce carré présente une parfaite symétrie logique. Malheureusement, il laisse en plan les formes utilisant un quantificateur moins [c'est entendu, je m'abstiendrai des expressions ambigues et ferai usage des termes classiques pour les deux types de quantificateurs: la *particulière* + ("quelque(s)", "un(e)", "au moins un(e)", etc.) +, et l'*universelle* − ("tout(e)", "tou(te)s", etc.).] Le carré fondamental illustre les formes particulières et leurs négations. Il y a deux formes universelles à examiner: "Tous les S sont P" (−S+P), notre a, et "Tous les S ne sont pas P" (−S−P), que nous appellerons e. Comment ces deux formes a and e peuvent-elles être caractérisées dans le carré? Il se trouve que, dans la large majorité des cas, nous pouvons nous en tenir aux définitions que l'on trouve dans notre logique du plus/moins et dire que a peut être définie comme A et e définie comme E. Lorsque tout se déroule normalement, nous pouvons tirer avantage de ces définitions en admettant les équivalences de a et A et de e et E. On pourrait illustrer ceci par le "Carré des Oppositions Normal".

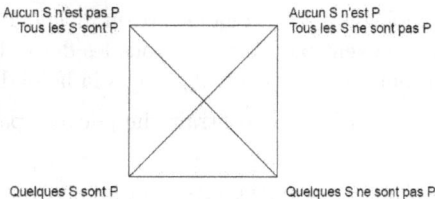

Mais les choses ne se déroulent pas toujours normalement. Notez que, dans le carré normal, les énoncés quantifiés particulièrement ont pour contradictoires non seulement leurs négations mais aussi les universelles qui sont logiquement équivalentes à ces négations. Rappelons les lois de non-contradiction et du tiers exclu. Nous pouvons les formuler ainsi:

LNC Un énoncé et son contradictoire ne peuvent pas être vrais tous les deux.

LTE Un énoncé ou son contradictoire est vrai.

Tant que tout se déroule normalement, $a = A$ et $e = E$, et nous pouvons dire que les contradictoires des énoncés I et O sont leurs négations. Mais nous avons fait allusion jusqu'ici au fait que les choses ne se déroulent pas toujours normalement. Préparons-nous au pire en formulant des lois qui gouvernent les énoncés universels (indépendamment de la façon dont elles pourraient être définies logiquement). Une loi que nous voudrions naturellement accepter, c'est le principe selon lequel il est impossible que des paires d'énoncés de forme "Quelques S sont P" et "Tous les S ne sont pas P" soient vraies toutes les deux (de même pour "Quelques S ne sont pas P" et "Tous les S sont P"). Appelons ceci la loi d'opposition quantifiée.

LOQ Un énoncé et son opposé quantifié ne peuvent pas être vrais tous les deux.

L'intuition nous incite à dire que, pour un sujet particulier quelconque, tout prédicat est tel que soit celui-ci soit sa négation est vrai de ce sujet. En termes de la portion de logique formelle que nous avons suggérée plus haut, cela veut dire que, pour tout énoncé quantifié particulièrement et un second énoncé exactement semblable à lui, excepté que le terme qui suit le qualificateur est la négation du terme qui suit le qualificateur du premier énoncé, un des deux énoncés est vrai. Nous voulons donc dire que soit "Quelques logiciens sont ennuyeux" soit "Quelques logiciens ne sont pas ennuyeux" est vrai (peut-être que les deux le sont). Appelons ceci la loi de subcontrariété.

LSC Un énoncé particulier ou son subcontraire est vrai.

Cette loi a un homologue. Nous voulons dire qu'un énoncé universel et son contraire logique ne peuvent pas être vrais tous les deux. La loi dit que les contraires logiques sont incompatibles. Appelons-la la loi d'incompatibilité.

LIC Un énoncé universel et son contraire ne peuvent pas être vrais tous les deux.

Toutes les lois formulées jusqu'ici gouvernent le carré des oppositions normal. Mais lorsque les choses tournent mal, la normalité n'est plus garantie. Les choses peuvent mal tourner de trois façons. Dans chacun de ces cas, le carré normal ne s'applique pas (au sens où une des lois n'est pas valable). De plus, la loi rejetée dans ces cas est le résultat de notre incapacité à définir les universelles (a et e) en termes des négations respectives (A and E).

3. L'import existentiel

Il va sans dire que les logiciens modernes ne formulent pas les énoncés des langages naturels de la façon que nous avons suggérée plus haut. Pour le meilleur ou le pire, ils standardisent les énoncés traditionnels de la façon suivante:

Ce "Carré des Oppositions Moderne" diffère des autres carrés à bien des égards. Notez par exemple que les quantificateurs, que nous avons présentés comme des parties inséparables (de pair avec les qualificateurs) de foncteurs de termes binaires scindés, ont été abandonnés au profit de nouveaux quantificateurs (qui sont conçus comme des fonctions appliquées à des énoncés ouverts entiers, liant des variables d'individus qui figurent dans ces énoncés). De plus, le quantificateur particulier a été remplacé par le quantificateur "existentiel". Cela veut dire que les nouvelles formes I et O ont un "import existentiel". Asserter un énoncé de ce genre, c'est faire une déclaration d'existence. En revanche, leurs contradictoires n'ont pas d'import existentiel. Cela veut dire que la relation traditionnelle de subalternation n'est pas valable pour le carré moderne.

Notre quantificateur particulier n'est pas existentiel. Mais nous devons nous demander ce qui arrive aux énoncés de forme "Quelques S sont P" et "Quelques S ne sont pas P" lorsqu'il n'y a pas de S. Il n'y a pas de satyres. Concédons ainsi au logicien moderne que "Quelques satyres sont des poètes" et "Quelques satyres ne sont pas des poètes" sont faux tous les deux. De façon plus générale, disons que lorsqu'il n'y a pas de S les formes I et O, "+S+P" et "+S−P", sont fausses toutes les deux. Mais le logicien moderne poursuit en disant que, dans ce genre de cas, les formes universelles correspondantes sont vraies (puisqu'elles ont été définies en effet comme les négations des formes I et O). En d'autres termes, le carré moderne est construit en utilisant les formes a, e, I et O, où a et e sont définies en termes de A et E. Nous ne suivrons pas le logicien moderne sur ce point. Disons que dans les cas où I et O sont fausses toutes les deux, leurs négations respectives sont vraies toutes les deux (en vertu de LNC et LTE); mais puisque LSC échoue, a et e ne peuvent plus être définies (en termes de A et E). Lorsque les choses tournent mal de cette façon, lorsqu'il n'y a rien qui satisfasse le terme quantifié particulièrement, aucune universelle correspondante ne peut même être définie (elles n'ont donc pas de valeur de vérité). Appelons *vides de sens* les cas de ce genre où LSC n'est pas valable.

4. L'indétermination

L'absence d'existence n'est pas la seule façon dont les choses peuvent mal tourner. LSC est mise en péril également par l'absence de détermination suffisante d'un sujet par un prédicat. Considérons à présent l'énoncé "Une bataille navale aura lieu dans le Golfe du Saint Laurent le 30 mars 2012". Pour le moment (et pour un certain temps), la valeur de vérité d'un tel énoncé est indéterminée. Supposons que nous le présentions dans un carré fondamental (à la position I). LNC et LTE seraient encore valables (ce qui est toujours le cas), mais pas LSC. Nous ne pourrions pas assigner (à présent) la vérité à "Une bataille navale aura lieu dans le Golfe du Saint Laurent le 30 mars 2012" ou à "Une bataille navale n'aura pas lieu dans le Golfe du Saint Laurent le 30 mars 2012". Lorsque les choses tournent mal de cette seconde façon, nous n'avons pas de moyens suffisants pour définir les universaux correspondants (a et e).

5. Le non-sens

Songez à l'orange que j'ai pressée pour mon jus de fruit du petit-déjeuner de ce matin. Il y a beaucoup de choses à dire à son sujet. Certaines sont vraies: "L'orange est fraîche", "Elle est sucrée", "Elle n'a pas de noyau". D'autres sont fausses: "L'orange est mauve", "Elle est aigre", "Elle n'est pas à moi". D'autres encore sont tout simplement des non-sens: "L'orange est courageuse", "Elle est analphabète", "Elle est un facteur premier de 36". De notre point de vue logique, on opère sur des paires de termes avec des foncteurs de termes binaires (scindés ou non) pour former des termes composés syntaxiquement plus complexes (dont certains sont des énoncés). Dans un grand nombre de cas (la plupart nous étant familiers dans notre discours ordinaire), la composition d'une paire de termes donne un terme sensé ("frais et orange", "Une orange est mauve", etc.). Mais la plupart des paires de termes ne "vont pas ensemble" du tout. La plupart de ces paires sont comparables à "orange"/"courageuse" (ou: "stylo"/"pleurer", "2"/"boire").

On peut raconter une très longue histoire sur ce type de non-sens, sur sa nature, sur ses conséquences fascinantes et de large portée à la fois pour la nature du langage et pour l'ontologie, ainsi que sur l'histoire des tentatives de rendre compte de tout ceci. Nous laisserons cette histoire de côté (vous serez heureux de l'apprendre) et accepterons simplement le fait que, parfois, nous pouvons former des énoncés qui constituent de purs et simples non-sens. Le non-sens est la troisième façon dont les choses peuvent mal tourner. Les énoncés dépourvus de sens ne sont pas vrais. Que dire de leurs cousins dans le carré?

Supposez que je dise "Quelques nombres sont des buveurs". Vous contesterez ceci naturellement. Vous ferez allusion au fait que les nombres ne sont pas le genre de choses que l'on peut qualifier de buveurs, que les termes "nombre" et "buveur" ne vont pas ensemble. Si vous vous montrez moins patients (ou moins enclins à philosopher), vous pourriez tout simplement nier ce que j'ai dit. Comment le feriez-vous? "Quelques nombres ne sont pas des buveurs"? "Ce n'est pas le cas que quelques nombres sont des buveurs"? Notez que le second énoncé est la négation logique (donc la contradictoire) de mon énoncé. Le premier est ambigu, en revanche. "Quelques nombres ne sont pas des buveurs" pourrait être paraphrasée soit en "Quelques nombres sont des non-buveurs", soit en "Ce n'est pas le cas que quelques nombres sont des buveurs" (le second énoncé). Mais "Quelques nombres sont des non-buveurs" est tout aussi vide de sens que mon énoncé initial. Il semble raisonnable de dire que vous pouvez nier ce que j'ai dit en assertant sa négation, et qu'une négation de ce genre est vraie. En conséquence, "Quelques nombres sont des buveurs" et "Quelques nombres sont des non-buveurs" sont faux tous les deux, alors que "Ce n'est pas le cas que quelques nombres sont des buveurs" et "Ce n'est pas le cas que quelques nombres sont des non-buveurs" sont vrais tous les deux. De plus, ces deux derniers énoncés sont paraphrasés plus simplement en "Aucun nombre n'est un buveur" et "Aucun nombre n'est un non-buveur". De façon générale, lorsque les termes "S" et "P" ne peuvent pas être composés sensément les formes "Quelques S sont P" et "Quelques S ne sont pas P" sont fausses toutes les deux, et leurs négations sont vraies toutes les deux (elles sont fausses sous leurs formes I et O et vraies sous leurs formes A et E); et les formes "Tous les S sont P" et "Tous les S ne sont pas P" (a et e) sont simplement indéfinies. LSC a donc échoué une fois de plus, alors que toutes les autres lois logiques sont valables ou ne s'appliquent pas.

6. Le carré hexagonal

Comme nous l'avons vu à présent, les formes a et e ont un pied-à-terre dans le carré normal tant que les choses sont ... normales, disons, tant que a et e peuvent être définies (et donc identifiées) à A et E. En bref, tant que rien ne tourne mal (qu'il n'y a pas d'absence d'existence, pas d'absence de détermination, pas de non-sens), a et e peuvent être définies et, par conséquent, LSC est valable. Mais supposons que nous ne sachions absolument pas si nos énoncés sont normaux ou non. Nous voulons être en mesure de présenter la totalité des relations logiques valables (ou pouvant l'être) dans la famille à six membres des énoncés de forme A, E, I, O, a et e. Ce qu'il faut, c'est une place dans le carré réservée aux deux dernières formes, sans rien préjuger en ce qui concerne leur définissabilité. De là l'"Hexagone des Oppositions".

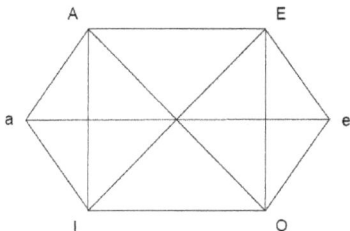

Dans les cas normaux, LNC, LTE, LOQ et LIC sont tous valables et peuvent être aperçues dans l'hexagone. Dans les cas normaux, a implique A, a implique I, e implique E, e implique O, A implique a, E implique e, et l'hexagone se confond simplement avec le carré normal. Lorsque les choses vont bien, LSC est valable, donc soit I soit O est vraie et soit A/a soit E/e est fausse. Dans ce genre de cas, tous les énoncés présentés sur un côté du carré sont vrais et tous les énoncés présentés sur l'autre sont faux. Dans les cas anormaux, a et e ne sont pas définis, l'hexagone ne se confond pas avec le carré, I et O sont faux tous les deux, A et E sont vrais tous les deux et, parce qu'ils sont indéfinis, a et e n'ont pas de valeur de vérité.

7. Mettre le carré au carré

Nous avons trouvé jusqu'ici une place dans le carré pour les formes A, E, I, O, a et e en transformant le carré en hexagone. Nous avons défini les deux premières de ces formes comme les négations des deux secondes; nous avons défini en effet les deux derniers en "introduisant", en distribuant le moins unaire initial de la négation dans chacun des deux premiers. Nous avons ainsi considéré I et O comme logiquement primitifs. Ce à quoi nous n'avons pas laissé de place dans l'hexagone, ce sont les négations des formes a et e. Notez que, compte tenu de nos formulations, I, O, a et e sont non-niées, alors que A et E sont des énoncés niés. Que dire des négations de a et e? La négation de a a la forme "$-(-S+P)$"; la négation de e a la forme "$-(-S-P)$". Appelons-les i et o, respectivement. Puisque a et e sont elles-mêmes non-primitives, i et o sont non-primitives également. De plus, puisque dans les cas non-normaux a et e sont indéfinis, il s'ensuit que i et o sont indéfinis également dans ces cas de figure. Dans les cas normaux, nous pouvons dire: A = a, E = e, I = i, et O = o. Nous pourrions présenter les huit membres de cette famille de formes étendue dans un "Carré des Oppositions mis au Carré".

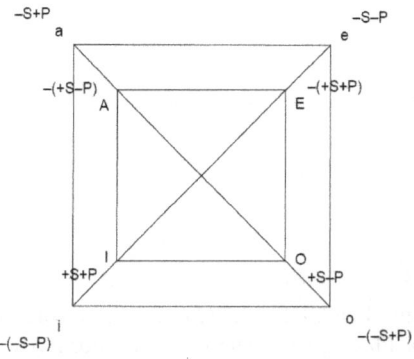

Une fois encore, tant que les choses vont bien le carré externe se confond simplement avec le carré interne.

Il se trouve que ce carré mis au carré nous permet d'apporter un éclairage sur certaines controverses. Par exemple, Parsons a proposé dans [13] une inférence invalide que l'on pourrait prouver en utilisant une version de la logique traditionnelle (celle proposée par Strawson dans [26]). La prémisse de l'inférence de Parson est "Aucun homme n'est une chimère" (qui est vraie), et l'énoncé inféré est "Quelque non-homme est une chimère" (qui est faux, puisqu'il y a des non-hommes mais il n'y a pas de chimères). Une application de la conversion à la prémisse donne "Aucune chimère n'est un homme". Puis de l'obversion de celle-ci on obtient "Toute chimère est un non-homme". Une application de la subalternation donne alors "Quelque chimère est un non-homme". Pour finir, cette dernière est convertie et produit "Quelque non-homme est une chimère". Il ne fait pas de doute que quelque chose a mal tourné dans cette preuve. Formulons-la dans notre système plus/moins.

1. $-(+H+C)$ prémisse
2. $-(+C+H)$ conversion sur 1
3. $-C-H$ obversion sur 2
4. $+C-H$ subalternation sur 3
5. $+(-H)+C$ conversion sur 4

Les choses semblent assez innocentes. Mais décrivons les formes de chaque étape. 1 et 2 ont des formes e; 4 et 5 sont de forme O. Il n'y a pas de raisons de rejeter les étapes menant de 1 à 2, de 3 à 4 ou de 4 à 5. Reste l'étape menant de 2 à 3. Or 3 est un énoncé de forme e; il est quantifié universellement et, comme nous l'avons vu, les énoncés quantifiés universellement (a et e) et leurs négations (i et o) sont définis dans les cas normaux et indéfinis dans les cas non-normaux. Puisque 3 est vide de sens (les chimères n'existant pas) il n'a pas de valeur de vérité, donc rien ne s'ensuit de 3, en particulier 4 (et 5, donc). La prémisse "Aucun homme n'est une chimère" est vraie parce qu'elle représente la négation de "Quelque homme est une chimère", qui est fausse. Mais "Quelque homme n'est pas une chimère" est fausse également, ce qui rend vide de sens la famille dans son ensemble. Donc le carré externe du carré mis au carré disparaît – la forme e (ligne 3), de même que a, i et o, a disparu de l'édifice.

Un bref commentaire de fin, avant de poursuivre: dans tous les cas de figure, soit le carré extérieur se confond avec le carré intérieur soit il disparaît, ce qui constitue peut-être la raison pour laquelle les logiciens y ont porté si peu d'attention.

8. Le coup de grâce

Toute notre discussion s'est concentrée jusqu'ici sur les termes supposés être généraux. Nos carrés et notre hexagone illustrent les relations logiques entre des familles d'énoncés dont les termes ne sont jamais singuliers. Mais un carré pourrait-il s'adapter aux énoncés singuliers? Supposons que le "S" de nos formes énonciatives puisse être interprété comme un terme singulier ("Socrate", "le premier homme sur Mars", etc.). Il semble raisonnable d'admettre que nous pouvons avoir à la fois des cas normaux et non-normaux même lorsque le terme sujet est singulier. Dans les cas anormaux, les choses peuvent tourner mal de chacune des trois façons déjà examinées. Dans le cas de "La plus jeune des filles du Prince Charles vit en Suisse", le terme singulier est vide de sens. Dans "Brad Pitt deviendra arboriculteur l'année prochaine", le sujet Brad Pitt est indéterminé par rapport à son futur. Et "Socrate est premier" est un non-sens.

Mais la question reste de savoir comment formuler logiquement des énoncés singuliers. Plus particulièrement, comment peuvent-ils être conçus avec la quantification (condition essentielle pour entrer dans le carré). Une réponse (encore une fois avec une histoire très longue, etc.) consiste à dire que les singulières peuvent être analysées logiquement comme des particulières, différant des particulières standards en ce qu'elles entraînent (pour des raisons sémantiques et non formelles) leurs universelles correspondantes. Cela voudrait dire par exemple que "Socrate est sage" a pour forme logique "+S+W", et qu'il entraîne "Tout Socrate est sage" (−S+W). Nous pouvons nous rappeler dans ce qui suit qu'un terme est considéré comme singulier en l'écrivant avec une lettre minuscule (nos énoncés singuliers deviennent ainsi "+s+P" et "−s+P"). Construisons donc un "Hexagone des Oppositions Singulier", pour commencer.

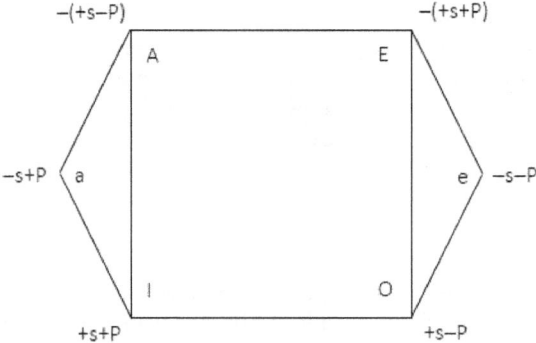

Comme à l'accoutumée, lorsque les choses vont mal a et e sont indéfinis, LSC échoue, etc. Mais lorsque les choses sont normales, non seulement a est défini comme A et e comme E (de sorte que a et A s'entraînent mutuellement, et de même pour e et E), mais I implique (sémantiquement) a de même que O implique e. Cela veut dire que l'hexagone singulier ne se confond pas simplement avec la contrepartie singulière d'un carré normal. Puisque a entraîne I et e entraîne O lorsqu'ils sont définis, et lorsque rien ne va de travers avec les singulières, l'hexagone simple se réduit à une simple ligne. De là la "Ligne Singulière des Oppositions".

A, I, a ——————————————————————————— E, O, e

Bibliographie

[1] Englebretsen, G., 1972, "Vacuousity", *Mind*, vol. 81, p. 273-275.

[2] Englebretsen, G., 1976, "The Square of Opposition", *Notre Dame Journal of Formal Logic*, vol. 17, p. 531-541.

[3] Englebretsen, G., 1980, "Singular Terms and the Syllogistic", *The New Scholasticism*, vol. 54, p. 68-74.

[4] Englebretsen, G., 1984a, "Quadratum Auctum", *Logique et Analyse*, vol. 107, p. 309-325.

[5] Englebretsen, G., 1984b, "Opposition", *Notre Dame Journal of Formal Logic*, vol. 25, p. 79-85.

[6] Englebretsen, G., 1986a, "Singular/General", *Notre Dame Journal of Formal Logic*, vol. 27, p. 104-107.

[7] Englebretsen, G., 1986b, "Singular/General", *Notre Dame Journal of Formal Logic*, vol. 27, p. 62-65.

[8] Englebretsen, G., 1988, "A Note on Leibniz's Wild Quantity Thesis", *Studia Leibnitiana*, vol. 20, p. 87-89.

[9] Englebretsen, G., 1996, *Something to Reckon With*, Ottawa: University of Ottawa Press.

[10] Englebretsen, G., 2006, *Bare Facts and Naked Truths*, Aldershot: Ashgate.

[11] Englebretsen, G., 2010, "Making Sense of Truth Makers", *Topoi*, vol. 29, p. 147-151.

[12] Oderberg, D. (ed), 2005, *The Old New Logic*, Cambridge, MA: MIT Press.

[13] Parson, T., 2006, "The Square of Opposition", *Stanford Encyclopedia of Philosophy*,
http://plato.stanford.edu/entries/square/ .

[14] Ryle, G., 1949, *The Concept of Mind*, London: HutchinsonŠs University Library.

[15] Ryle, G., 1959, "Categories", *Logic and Language*, 2e édition, A. Flew (éd.), Oxford: Blackwell.

[16] Sommers, F., 1959, "The Ordinary Language Tree", *Mind*, vol. 68, p. 160-185.

[17] Sommers, F., 1963, "Types and Ontology", *Philosophical Review*, vol. 72, p. 327-363.

[18] Sommers, F., 1965, "Predicability", *Philosophy in America*, M. Black (éd.), Ithaca: Cornell University Press.

[19] Sommers, F., 1967, "On a Fregean Dogma", *Problems in the Philosophy of Mathematics*, I. Lakatos (éd.), Amsterdam: North-Holland.

[20] Sommers, F., 1969, "Do We Need Identity?", *Journal of Philosophy*, vol. 66, p. 499-504.

[21] Sommers, F., 1982, *The Logic of Natural Language*, Oxford: Clarendon Press.

[22] Sommers, F., 1987, "Truth and Existence", *The New Syllogistic*, G. Englebretsen (éd.), New York: Peter Lang.

[23] Sommers, F., 1993, "The World, the Facts, and Primary Logic", *Notre Dame Journal of Formal Logic*, vol. 34, p. 169-182.

[24] Sommers, F., 1996, "Existence and Correspondence to Facts", *Formal Ontology*, R. Poli & P. Simons (éd.), Dordrecht: Kluwer.

[25] Sommers, F. & G. Englebretsen, 2000, *An Invitation to Formal Reasoning*, Aldershot: Ashgate.

[26] Strawson, P. F., 1952, *Introduction to Logical Theory*, London: Methuen.

George ENGLEBRETSEN
Bishop's University, Canada

Chose, Concept et Molécularisme : autour de la proposition des *Principles* de Russell

PHILIPPE GAC

Au tout début du XXe siècle, Russell considère que "la logique n'est pas intéressée par les mots mais par ce qu'ils représentent". Un groupe nominal désigne en principe un objet, éventuellement abstrait. Un énoncé, c'est-à-dire le texte constitué de signes dans une langue donnée, exprime une "proposition", un objet platonicien qui a une existence indépendante des contingences linguistiques. Russell, dans ses *Principles of Mathematics* [12], tente de définir ces objets platoniciens, simples ou complexes, en corrélation avec le lexique (anglais), et leurs règles d'assemblage, calquées de fait sur la syntaxe (anglaise). La description d'un tel système emploiera un certain langage idéalisé censé refléter exactement la réalité logique. Là où Frege n'hésite pas à recourir à des diagrammes ou à des idéogrammes, Russell cherche plutôt à enrégimenter le langage naturel, d'autant que "la grammaire, globalement, [lui] semble nous rapprocher davantage d'une logique correcte que ce que pensent les philosophes" (§46a[1]). Ainsi, "la grammaire sera [dans l'ouvrage] certes pas notre maître, mais néanmoins notre guide" pour comprendre et exprimer la réalité logico-sémantique sous-jacente, conduisant à "ontologiser la grammaire" [10].

Le projet s'engage de manière cohérente dans plusieurs choix ontologiques, dont un certain atomisme[2], sans doute métaphysique ou sémantique, mais pas encore "logique", ni même vraiment épistémique. Des entités atomiques, les "termes" (ou "entités"), sont primitifs, postulés, "indéfinissables" ; il n'est pas forcément impossible de les décrire ou de les désigner mais cela ne suffirait pas à les fonder logiquement ; leurs propriétés élémentaires doivent aussi être postulées. Ces entités permettent de construire d'autres objets, et ce, récursivement. Les objets complexes sont tous les objets

---

[1] Tous les numéros de paragraphe font référence aux *Principles of Mathematics* [12] ; l'éventuelle lettre suffixée numérote l'alinéa. Le texte cité a été traduit par l'auteur de l'article.

[2] Russell emploiera le vocabulaire de l'atome seulement vers 1913, et dans un sens différent.

mathématiques classiques (ensemble, tuple[3], ...) mais aussi les propositions, dénotations (exprimées par un groupe nominal descriptif ou une expression mathématique) et d'autres, relativement inédits (disjonction ...). L'un des soucis de Russell, l'"analyse", est de ramener un objet complexe à ses constituants, organisés dans un champ structural certes ultimement "indéfinissable" mais qu'il s'efforcera autant que possible de réduire et simplifier. Il échappe ainsi aux objections traditionnelles des holistes : "bien que l'analyse nous livre la vérité, et rien que la vérité, elle ne peut jamais nous livrer toute la vérité, du fait de certaines relations ultimes et indéfinissables" (§§138, 137), de la même façon que l'analyse chimique ne décrit pas forcément les conformations des molécules. Inversement, des combinaisons d'entités constituent des "objets" ou même induisent de nouvelles entités, des "unités". Russell distingue ainsi entre la classe multiple (la pluralité des humains) et la classe collective ou "classe en tant qu'une" (l'humanité en tant que tout), encapsulation de la première en une unité, une entité de plein droit qui peut remplacer un atome dans toute proposition, la rendant au pire fausse.

On s'attachera ici[4] à examiner des difficultés fondamentales liées à cette conception, déjà nombreuses et intéressantes, même en se limitant aux énoncés simples, positifs, déclaratifs, singuliers et intemporels de certaines langues. La question centrale, inédite, est la possibilité même du molécularisme de la proposition des *Principles*.

1 La proposition et ses constituants.

Selon Russell, la proposition "*Socrate est humain* contient un terme et un prédicat" (§57b ; §53) : Socrate lui-même en est un "constituant"[5]; *Napoléon* et *Bonaparte* sont donc une seule et même entité et cela évite

---

[3]Un tuple, ou n-uplet, est un ensemble ordonné. Pour $n=2$, un couple, ou paire ordonnée, est noté (a, b) et diffère de l'ensemble $\{a, b\}$, qui est une paire (non ordonnée), en ce que (a, b) et (b, a) sont deux couples différents tandis que $\{a, b\}$ et $\{b, a\}$ sont le même ensemble, et que $\{a, a\}$ se réduit au singleton $\{a\}$ alors que (a,a) est un couple de plein droit (mathématique). Pour $n=3$, un triplet ordonné est noté (a, b, c).

[4]L'espace imparti a imposé de renoncer à préciser l'ancrage historique (Platon, Leibniz, Bolzano, Frege, Bradley, Meinong, Moore ...) ou les références non indispensables à la compréhension.

[5]Il ne faut pas pour autant imaginer Socrate ressuscité allant d'une proposition à l'autre le temps qu'elle soit affirmée : les objets russelliens doivent être compris comme des êtres quadrimensionnels dont nous ne percevons que des tranches tridimensionnelles. Ils participent à des propositions éternelles et leur trajectoire spatio-temporelle n'est qu'une propriété parmi les autres, autorisant, à la limite, des objets fictifs de trajectoire spatio-temporelle vide. La distinction entre objets concrets (*Socrate*) et abstraits (*humain*) s'estompe alors, et leur combinaison est encore, d'une certaine façon, dans le même monde. A priori, ces êtres peuvent aussi apparaître en de multiples occurrences d'une proposition, à l'instar des objets mathématiques.

les problèmes inhérents au langage, de l'arbitraire et de la multiplicité des dénominations. Le concept *humain* est bien un terme, et l'entité est présente dans la proposition, mais au lieu d'être utilisée "comme terme" à l'instar de Socrate, et comme dans *"humain" est un concept*, elle a en outre un rôle déterminant dans la structuration de la proposition. Il ne correspond en effet aucune entité à la copule *est* et les deux entités se fondent dans "l'unité de la proposition", indicible et mystérieuse ; celle-ci est induite par des "relations externes" qui relient les deux entités, *externes* au sens où elles ne sont pas des propriétés de l'un des éléments, ni même de leur couple ou d'une entité extérieure[6]. A priori, ce champ structural qui est censé organiser les entités en une proposition est une vertu unitive guère plus extravagante que le champ gravitationnel newtonien ou les trois autres forces qui maintiennent l'atome physique.

Russell est réticent à introduire de façon systématique des entités ad hoc supplémentaires, surtout lorsque la difficulté que les nouvelles entités devaient permettre de contourner serait seulement déplacée ; autant, par conséquent, admettre la structure de départ. Un exemple édifiant est l'acceptation des relations externes et la réfutation des succédanés de relation (§§212-216), de "la théorie moniste qui maintient que toute proposition relationnelle aRb doit être ramenée à une proposition concernant le tout composé de a et b", et de la "théorie monadiste" qui édicte "que la proposition "A est plus grand que B" doit être analysable en deux propositions, l'une attribuant un adjectif à A [être plus grand que B], l'autre un adjectif à B [être moins grand que A]" (§214b ; §94b). Pour commencer, dans le cas d'une relation asymétrique (comme $a<b$), la théorie moniste doit être amendée en considérant un tout ordonné[7], donc en utilisant déjà une relation asymétrique : au mieux, on ramène les ordres à un ordre unique, et cela appelle d'autres conventions.

À partir de deux entités dont une seule est un prédicat, la proposition "est

---

[6]Russell veille à éviter la récursion infinie : ce qui relie les entités *Socrate* et *humain* ne peut être seulement une entité car le problème se poserait à nouveau de relier cette dernière aux deux premières. Et une entité qu'exprimerait la copule lui semble superflue, qui mettrait *Socrate* et *humain* dans des positions équivalentes, ce qui n'est clairement pas le cas dans la langue guide.

[7]C'est-à-dire ici, un couple et non pas un ensemble à deux éléments. En effet, $a<b$ ne se ramène pas à $<(\{a, b\})$ mais seulement à $<((a, b))$ et écrire (a, b), c'est déjà ordonner (les occurrences) a et b. Pour passer du tuple à l'ensemble, il suffit d'oublier l'ordre. Inversement, les tuples peuvent être codés avec des ensembles de diverses manières, dont celle de Kuratowski (1921), $(a, b) = \{\{a\}, \{a, b\}\}$. Ensuite, on définit aisément $(a, b, c) = (a, (b, c))$, mais ce n'est qu'une possibilité parmi d'autres, au choix donc arbitraire. Le gros défaut de ces constructions est qu'on obtient tout au plus des représentations, conventionnelles, du couple, non le couple lui-même : la variante inverse $\{b, \{a, b\}\}$ est tout autant légitime.

définie dès que les constituants sont donnés" (on y reviendra). Les choses se compliquent avec l'arrivée d'un deuxième terme (donc d'une troisième entité) car la structure n'est plus univoque : *Socrate, Platon* et *écouter*, au niveau logique, s'arrangent aussi bien en *Socrate écoute Platon* qu'en *Platon écoute Socrate*. Comme en chimie avec les isomères (des molécules constituées des mêmes atomes arrangés différemment), on ne peut reconstituer la molécule à partir de la liste des atomes qui la composent que dans les cas simples. Cet exemple montre même que l'isomérie peut ici commencer à partir de seulement trois atomes car les points d'attache d'une relation sont rarement équivalents, contrairement à ceux de la liaison chimique. En effet, pour un atome bivalent, lié à deux autres atomes distincts, disons l'oxygène à l'hydrogène et au chlore dans l'acide hypochloreux H-O-Cl, cela n'a aucun sens de distinguer entre H-O-Cl et Cl-O-H puisque ce serait la même molécule ayant pivoté d'un demi-tour.

En chimie, seules interviennent les positions relatives des atomes satellites les uns par rapport aux autres. Les positions absolues présupposent déjà une certaine structure d'espace ; une position relative revient à identifier les positions absolues superposables via une rotation dans cet espace. Si on reste dans le plan (à deux dimensions), l'isomérie apparaît avec le troisième satellite car ABC autour de O (dans le sens des aiguilles d'une montre) n'est pas superposable modulo une rotation à CBA (dans le sens inverse). Par contre, dans notre espace à trois dimensions, CBA est le retournement de ABC par rapport au plan contenant et ce n'est qu'à partir de quatre atomes périphériques différents que les configurations ne sont plus identiques, donnant en principe lieu à deux énantiomères, des isomères qui sont l'image dans un miroir l'un de l'autre, comme la main gauche et la main droite. Par exemple, un atome de carbone entouré de quatre atomes différents (brome, chlore, fluor, iode) donne lieu à deux énantiomères (de bromochlorofluoroiodométhane, CBrClFI, à se représenter dans l'espace tridimensionnel, cf. figure 1).

Figure 1. la molécule du bromochlorofluoroiodométhane (CBrClFI)

Cela montre l'importance du substrat géométrique dans lequel évolue l'assemblage : dans un espace à quatre dimensions, de nouvelles rotations apparaissent qui confondraient ces deux isomères[8]. À l'inverse, une chimie unidimensionnelle ressemblerait davantage aux constructions langagières : H-O-Cl serait alors discerné de la molécule symétrique Cl-O-H, et cela pourrait être mis en évidence physiquement si l'axe était orienté, par exemple, par un champ électrique, tout comme la suite des mots est orientée par le sens conventionnel de lecture. L'analogie s'arrête là car le langage autorise des constructions arborescentes difficilement transposables dans une chimie strictement linéaire. En fait, cette isomérie géométrique n'est pas celle qui semble pertinente pour la sémantique.

Dans l'exemple précédent, *Socrate écoute Platon*, il faut connecter *Socrate* et *Platon* à *écouter* : on postule donc que *écouter* a deux emplacements, ou paramètres, *écouté* et *écoutant*, par lesquels, respectivement, les arguments[9] *Socrate* et *Platon* sont liés à *écouter*. On supposera ici que le mode de connexion est alors parfaitement déterminé. Le verbe asymétrique évoque alors plutôt les troncs moléculaires de la chimie dans l'isomérie structurale, où les emplacements autour d'une molécule constituée de plusieurs atomes sont différenciés selon l'atome auquel se rattache le satellite. Par exemple, le tronc -CH_2-O- fournit deux points d'attache non équivalents : selon qu'on accole un atome de chlore à l'atome de carbone (à gauche) ou à celui d'oxygène (à droite), en complétant par un atome d'hydrogène, on obtient le chlorométhanol ou l'hypochlorite de méthyle, de même formule moléculaire (CH_3ClO) mais aux propriétés chimiques très différentes.

Enfin, les atomes périphériques peuvent se voir substituer des molécules (de valences équivalentes), tout comme dans le langage un nom propre peut être remplacé par un groupe nominal complexe ; le CBrClFI devient alors, par exemple, l'acide lactique $CHOHCH_3COOH$, avec similairement deux énantiomères, distinguables optiquement et biologiquement. Il ne peut y avoir isomérie que si les deux arguments sont interchangeables : *Socrate regarde Platon* est isomérique mais non *Socrate est humain* car Russell stipule que les termes sont substituables mais non les prédicats (cf. infra), lesquels n'ont justement pas la même valence que les termes.

En sémantique comme en chimie, les forces qui assemblent et structurent

---

[8] De façon générale, dans un espace à n dimensions, les assemblages sont forcément indifférenciés jusqu'à n satellites. Quant à l'espace logique, il semble naturel de lui supposer comme une infinité de dimensions, ce qui conduit à confondre tous les isomères précédents.

[9] Pour la clarté de l'exposé, on reprend ici le vocabulaire de l'informatique, largement postérieur au texte de Russell, lequel emploie, dans le cadre d'une fonction propositionnelle, *variable* et *constante* (§41), idoines comme qualificatifs mais trop généraux et déjà surexploités.

les éléments sont ici présupposées, postulées et sont extérieures à la théorie ; les relations externes et les liaisons chimiques n'ont qu'une valeur descriptive. Cependant, il faudrait préciser ce qui différencie les structures des propositions, et joue le rôle de la position spatiale relative en chimie, cela, bien sûr, sans faire un appel implicite à la forme linguistique[10]. Le langage marque la correspondance paramètre-argument par trois moyens non exclusifs, présents à divers degrés dans chaque langue : la position sur l'axe syntagmatique, la flexion (déclinaison) des noms ou la présence de prépositions. Il se trouve qu'aucun de ces moyens n'est transposable au niveau de la proposition avec l'exigence de naturalité, c'est-à-dire sans instaurer des conventions arbitraires. Or, les relations, objets platoniciens qui interviennent dans des propositions censées émaner directement de la réalité, ne peuvent être soumises à de telles conventions arbitraires voire aléatoires.

Premièrement, la position revient à attribuer des numéros d'ordre aux arguments respectivement à ceux des paramètres, supposant qu'aucun n'est omis. Il faut convenir artificiellement que, par exemple, l'*écoutant* est placé en première position, ce qui est peu satisfaisant dans une entreprise qui cherche à s'abstraire de la convention linguistique, d'autant que ce numéro dépendra fatalement de la relation. En effet, dans les langages informatiques aussi, lorsque l'ordre des arguments est significatif, l'ordre des paramètres est souvent arbitraire et parfois contradictoire d'une fonction à l'autre : au delà de quelques conventions uniformes dans des cas simples et généraux, il n'est simplement pas possible d'adopter un ordre cohérent, alors même que les fonctions sont les créatures du programmeur.

Deuxièmement, la flexion revient à démultiplier les entités, *Platon* en *Platonant* (un avatar de Platon pour le sujet) et *Platoné*, ou plutôt, si on veut éviter l'arbitraire de la relation entre de telles entités élémentaires, il faut admettre des entités dérivées *Platon-ant* (Platon comme sujet) et *Platon-é*, telle une unité englobant *Platon* et *ant* (Platon vêtu d'une toge de sujet). Cela conduit certes à des structures différenciées (*Socrate-ant*) - *écouter* - (*Platon-é*) contre (*Socrate-é*) - *écouter* - (*Platon-ant*) mais le choix de *ant* comme marqueur de l'argument sujet de *écouter* est arbitraire et, comme pour les ordres des arguments, il est impossible d'étiqueter ainsi les paramètres de façon cohérente sur l'ensemble des relations : comment devrait-on marquer le sujet de *enseigner, parler, voir* ? Platon devrait

---

[10]Notamment, l'ordre conventionnel de la juxtaposition syntagmatique, fondée sur le substrat spatial de la ligne d'écriture ou temporel du flux sonore. Dans la quasi-totalité des langues et des langages informatiques, les énoncés sont constitués par la suite des mots ordonnée sur un axe avant/après (gauche/droite dans les langues européennes). Divers dispositifs permettent une certaine liberté mais il n'en reste pas moins que, même en latin, l'ordre de certains mots au moins est significatif.

revêtir à la fois une toge de sujet lorsqu'il écoute Socrate et une toge de destinataire dès que Socrate lui parle, alors que sa toge de sujet le désignerait comme celui qui parle. La seule échappatoire est de spécialiser les étiquettes en *Platon-écoutant* et *Platon-parlé* (Platon est vêtu d'une multitude de toges mais au moins elles ne sont plus contradictoires) mais il y a maintenant une relation arbitraire entre *écoutant*, *écouté* et *écouter* ... Si on considère des étiquettes complexes *écouter-ant* et *écouter-é*, on obtient une proposition (*Platon - (écouter-ant)) - écouter - (Socrate - (écouter-é)*). On est alors parvenu à doubler la difficulté initiale sans la dépasser car il faut maintenant expliquer la combinaison *Platon - (écouter-ant)*, ce qui est tout aussi difficile que d'expliquer directement *Platon - écouter - Socrate*, et le lien entre *écouter* et *ant*, qui n'est pas celui d'une prédication, reste à élucider[11]. De surcroît, les étiquettes *écouter-ant* et *écouter-é* sont à peine moins conventionnelles que *écouter-1* et *écouter-2*.

Troisièmement, enfin, l'analogue des prépositions consisterait à postuler des entités supplémentaires qui s'intercaleraient entre la relation et ses arguments (sauf éventuellement l'un d'entre eux, le sujet), distinguant ainsi *Socrate - ant - écouter - é - Platon* de son converse *Platon - ant - écouter - é - Socrate*. Déjà, on est confronté comme précédemment à l'arbitraire des connecteurs (- *ant* -). Si l'on est en chimie, la structure doit respecter les valences de ses constituants : *Socrate* et *Platon*, monovalents, sont forcément en bout de chaîne et les autres, bivalents, au milieu. Pour imposer l'ordre (admettre - *ant - écouter - é* - mais interdire - *ant - é - écouter* -), il faut postuler une différence entre les connecteurs et les autres atomes (par exemple, comme entre raccords mâles et femelles), mais cela ne suffit pas à dispenser de postuler une certaine forme d'ordre qui permette de distinguer différents tuples à partir d'un ensemble d'atomes. Autant décider dès le départ que -*écouter* - est orienté ; il n'y avait de toute façon vraiment aucune raison de supposer que le verbe était symétrique comme un atome. Le molécularisme sémantique admettra donc que les emplacements argumentaux des relations sont différenciés et ce d'une manière "indéfinissable".

2 Le prédicat et la chose.

Dans le langage naturel, du moins dans certaines langues, la distinction entre prédiquant et prédiqué est marquée, dans les phrases simples, par l'opposition entre adjectif et verbe d'une part, et nom d'autre part. L'analogie chimique montre là ses limites : chimiquement parlant, les noms sont monovalents, les adjectifs bivalents et les verbes, plurivalents ; ce n'est toute-

---

[11] Ce raisonnement est fortement suggéré par la réfutation du monadisme évoqué plus haut (§214b) dont l'assemblage peut se schématiser A - (*plus_grand* $\sim$ B), où le second lien ($\sim$), la "référence" à B, non prédicatif, est "inintelligible" (§54).

fois qu'une condition nécessaire car on ne peut former une proposition avec un adjectif et deux noms qui serait l'analogue de la molécule d'eau (§422a). On préfère donc dire que le nom est niladique, se suffisant à lui-même (saturé, disait Frege), l'adjectif, monadique, puisque qualifiant un nom (éventuellement déjà qualifié par un ou plusieurs adjectifs) et le verbe polyadique, prenant un sujet et des compléments.

Parmi les constituants d'une proposition, Russell oppose les "prédicats" aux "sujets", ou "termes", qui partagent "la propriété caractéristique que l'un quelconque d'entre eux peut être remplacé par n'importe quelle autre entité sans qu'on cesse d'avoir une proposition" (§48b). Si on remplace *Socrate* par n'importe quel autre terme, on obtient au pire une proposition fausse : *2 semble humain* est banalement faux. En revanche, si on remplace un prédicat par une entité non prédicative, on n'obtient pas de proposition : *Socrate semble Athènes*[12] est ainsi un assemblage impossible, dénué de sens, qui n'est ni vrai ni faux, puisqu'il n'est pas une proposition (et vice versa!).

Plus précisément, *Socrate semble Socrate* n'exprime pas une proposition tandis que *(le concept) "humain" semble humain*, quoique faux, est pour Russell bien formé, où l'on fait successivement mention puis usage du concept. Une alternative se présentait : soit *"humain"* et *humain* désignent deux entités distinctes quoique éventuellement dérivées l'une de l'autre ou d'une entité commune, soit c'est une seule et même entité employée de deux manières différentes. Russell érige la seconde possibilité en principe : "tout constituant de toute proposition doit pouvoir être pris comme sujet logique" (§52b), tel quel, et ce, simplement de par sa place dans la proposition, ce qu'il appellera par la suite le "mode d'occurrence" de l'entité dans la proposition. De *Socrate est une entité* à *(le concept) "humain" est une entité*, il suffit de remplacer l'entité *Socrate* par l'entité *"humain"*, laquelle n'est rien d'autre que l'entité *humain*.

Un "concept" prédicatif (*class-concept*) est ainsi une entité qui *peut* être employée comme prédicat et admet de ce fait un emploi autre que celui comme terme. Au contraire, "Socrate est une chose car Socrate ne peut jamais intervenir autrement qu'en tant que terme dans une proposition et n'est pas susceptible de ce curieux emploi double qui est en cause avec *humain*" (§48b). Autrement dit, certaines entités ont la propriété intrinsèque d'être utilisables comme prédicat ; l'utilisation effective est en revanche une propriété extrinsèque mise en œuvre au sein d'une proposition.

Ces emplois sont distingués d'emblée dans le langage, au niveau du lexique : "*humain* et *humanité* [au sens de la propriété d'être humain] désignent précisément le même concept [*humain*], l'un ou l'autre de ces mots étant

---

[12]On évite le verbe *être*, trop équivoque ; *Socrate est Athènes* est tout aussi incorrect lorsque *être* est pris strictement en son sens de copule.

employé, selon le type des relations que ce concept entretient avec les autres constituants de la proposition dans laquelle il intervient" (§46). Les restructurations de la phrase et de la proposition se correspondent étroitement : "la proposition *L'humanité appartient à Socrate*, qui est [logiquement] équivalente à *Socrate est humain*, est une assertion à propos du concept d'humanité", où une relation *appartient* enrôle deux termes arguments. Par contre, la seconde proposition, qui est "distincte de la première, n'établit pas une relation entre Socrate et l'humanité puisque cela ferait intervenir *humain* comme terme" (§57) or elle implique l'entité *humain* seulement comme prédicat.

Dans le cas de *Socrate est humain*, il n'y a qu'une seule façon de relier les entités, celle qui fait de *humain* le prédicat qui s'applique à *Socrate*, car cela n'aurait a priori pas de sens de prédiquer similairement *Socrate* de *humain*. La structuration de la proposition, si proposition il y a, est immanente (elle ne dépend que des propriétés de ses constituants) et la structure peut être vue comme émergeant spontanément, sous la contrainte de fournir une proposition. Ces distinctions de type, liées aux parties du discours des grammairiens, contribuent, comme avec des pièces préfabriquées, à précontraindre la structure de la proposition, mais elles n'y suffisent pas toujours. Si Socrate est remplacé par un prédicat, on obtient potentiellement deux propositions isomères, par exemple, *"essentiel" est humain* et *"humain" est essentiel*, éventuellement fausses, mais non dénuées de sens.

Par ailleurs, cette ambivalence métaphysique sujet / prédicat ne recoupe pas du tout une ambivalence langagière tout aussi systématique. Diverses langues distinguent le prédicat du concept mais utilisent le nom conceptuel en divers sens : la propriété elle-même simplement en tant qu'elle est vraie (*L'humanitude de Socrate est certaine*), sa façon (*L'humanité de Socrate est exemplaire*), voire son avatar (ou *trope*, suivant un tropisme littéraire), concrétisation de la propriété comme partie d'un objet concret (*L'humanité de Socrate s'est éteinte avec lui*). En outre, le nom désigne ici également la collectivité des objets qui ont la propriété (*Socrate appartient à l'humanité*, à ne pas confondre avec *L'humanité appartient à Socrate*!). Afin de s'assurer de la justesse de l'emploi, on testera le remplacement du nominalisé par d'autres mots ou expressions (*caractère humain, propriété d'être humain*, etc.), éventuellement néologiques (*humanitude*). Tous ces objets que peut désigner le mot du concept, auxquels le prédicat est lié indépendamment du langage, dépendent aussi du sujet prédiqué, voire sont intriqués avec lui, sauf le concept en soi, invariant par définition ; celui-ci était donc le seul susceptible d'être identifié au prédicat présumé invariant.

Mais cela ne dit pas quelles entités sont des concepts-prédicats et Russell n'explique pas vraiment pourquoi *Socrate* n'en est pas un. Il en vient notamment à cette définition plus ou moins circulaire : "ce qui caractérise un concept prédicatif, par rapport aux termes en général, est que "x est un a" est une fonction propositionnelle si, et seulement si, a est un concept prédicatif" (§58c). Dans le cas contraire, "on n'a pas une proposition fausse, mais simplement pas de proposition du tout, quelle que soit la valeur qu'on donne à x" car, pour que deux entités puissent constituer une proposition, il faut obligatoirement que l'une tienne le rôle de prédicat. En somme, *Socrate* n'est pas un prédicat car il ne peut pas être employé comme prédicat afin de constituer une proposition avec une entité qui n'en est pas un non plus...

Or on peut imaginer des emplois de *Socrate* comme prédicat, l'entité prédiquant prenant au besoin la forme linguistique *socratien*. Hypothèse la plus naturelle, *socratien* serait la propriété d'être identique à *Socrate*, un prédicat qui caractérise *Socrate* de manière unicitaire et générique, et dont le contenu conceptuel se réduit au fait d'appartenir à cette extension singleton. Un tel prédicat décrit son extension de façon immanente (ne faisant appel à aucune propriété extérieure) et rigide (quelles que soient les circonstances, le seul *socratien* est Socrate)[13]. Dès lors qu'on peut identifier l'extension (*Socrate*) au prédicat de l'extension qui lui est ainsi canoniquement[14] associé (*socratien*) et est cette fois unique, on peut envisager de défendre l'idée que *Socrate* et *socratien* sont une seule et même entité, différant seulement dans leurs emplois.

Cependant, le rapport entre *Socrate* et *socratien* n'est pas forcément le même qu'entre *humanitude* et *humain*. Russell note que si "parmi les substantifs, certains sont dérivés d'adjectifs ou de verbes, comme *humanité* de *humain*, d'autres tels que les noms propres, ou l'espace, le temps et la matière ne sont pas dérivatifs mais apparaissent de manière primaire comme substantifs" (§46b). Certes l'adjectif *socratien* est dérivé lexicalement du nom de son (unique) occurrence mais cela n'est pas propre aux noms propres (cf. *défaut, défectueux* ; *chance, chanceux*) même si c'est plus communément l'inverse pour les noms communs (cf. tableau ci-après). Autre obstacle, *Socrate* n'est pas un concept, comme *humanitude*, mais une entité

---

[13]On trouve une bonne comparaison en géométrie : un ensemble de points peut être donné comme décrit par une équation sur les coordonnées, ou bien par une réunion / intersection de diverses figures plus ou moins remarquables, mais aussi par l'énumération de ses points, un par un. Il y aurait ainsi pour chaque extension ce qu'on pourrait appeler *le* concept caractéristique. C'est presque la fonction caractéristique du singleton $\{Socrate\}$, qu'on écrirait en λ-calcul : $socratien := \lambda x \cdot (x = Socrate)$, défini quel que soit le type de x.

[14]On dispose donc d'une injection canonique dans les concepts, c'est-à-dire une fonction, définie pour toute extension (elle associe un unique concept à chaque extension), injective (chaque concept associé l'est à une unique extension), de façon générique (sans faire appel à quelque spécificité de quelque valeur, dès lors que l'identité est définie) et de façon naturelle (les opérations du domaine de départ correspondent à celles du domaine d'arrivée).

spatio-temporelle, qu'on aurait tendance à rapprocher de *l'humanité*, ensemble des individus humains concrets, sinon que être socratien, c'est être identique à *Socrate*, mais, être humain, ce n'est certainement pas être égal à l'ensemble des humains, fût-ce "la classe en tant qu'une" (§§ 70, 79), mais plutôt, et tout au mieux, à leur disjonction, au sens technique qu'en donne Russell (§59). Cela n'est toutefois pas rédhibitoire car l'emploi du nom de la propriété pour quelque forme de concrétisation est aussi contraint de façon arbitraire et irrégulière et les divers emplois (propriété, occurrence ; adjectif ...) ne se recouvrent souvent que (très) médiocrement.

Quoi qu'il en soit, le prédicat *socratien* doit bien admettre un substantif spécifique *socratitude*, suivant le paradigme *Socrate est socratien / La socratitude appartient à Socrate*. On aboutit alors à un tableau qui semble conforter la légitimité du prédicat *socratien* (Figure 2) :

concept	prédicat	concrétisation
humanitude	*humain*	*l'humanité* = les individus humains
socratitude	*socratien*	*Socrate* = l'individu Socrate
défectuosité	*défectueux*	*un défaut* = une partie défectueuse
perversité	*pervers*	*une perversion* = un trait pervers
similitude	*similaire*	*une similarité* = une partie similaire
infinitude	*infini*	*Infini*
antillitude	*antillais*	*les Antilles* = les îles antillaises

Figure 2. la légitimité du prédicat "socratien"

En revanche, il apparaît que *socratien* devrait être identifié avec *socratitude* plutôt qu'avec *Socrate*, du moins si l'on admet, à la suite de Russell, que *humanitude* = *humain*. Alors, *Socrate* ≠ *socratien*, à moins bien sûr que *Socrate* = *socratien* = *socratitude*, et *un défaut* = *défectueux* = *défectuosité*, ce que Russell refuse catégoriquement car *un défaut* est aussi un concept, mais d'un type radicalement distinct.

3 Le nom propre et la description définie

Russell distingue entre le concept prédicatif *humain* qui peut être prédiqué de certaines entités et le concept dénotatif (*denoting concept*), *les humains*, qui, par essence, fait référence à une ou plusieurs entités, sa dénotation, via une description qui comporte des concepts prédicatifs, mais à laquelle il ne se réduit absolument pas : "*humain* et *humanité* diffèrent seulement grammaticalement", non en tant qu'entités logiques, mais les concepts "*homme*, *un homme* sont authentiquement distincts les uns des autres. Il y a, reliés à tout prédicat, une grande variété de concepts liés qu'il est important de distinguer" (§§ 58a, 48b, 489). Une "description définie", c'est-à-dire un groupe nominal descriptif débutant par un article défini, exprime une signification (*meaning*) qui est le concept dénotatif défini lui-même (*le maître de Platon*), lequel dénote sa dénotation (Socrate). Ainsi, les groupes nominaux dénotatifs "ont deux faces, un concept dénotatif et un objet dénoté".

Par définition, "un concept dénote lorsque, s'il intervient dans une proposition, cette proposition porte non pas sur le concept [de *maître de Platon*] mais sur un terme [*Socrate* lui-même] qui lui est relié d'une certaine manière par la relation de dénotation" (§§ 51, 56). Ainsi, *Socrate est humain* contient *Socrate* mais dans *Le maître de Platon est humain*, le "concept dénotatif prend sa place", et la proposition est alors seulement "à propos de" *Socrate*, qui n'en est pas un "constituant". Le concept dénotatif défini est très particulier en ce que le groupe nominal correspondant peut souvent être remplacé par un nom propre de l'entité qu'il dénote, ce qui revient, au niveau logique, dans la proposition, à substituer au concept dénotatif l'entité dénotée elle-même : si le concept *le maître de Platon* dénote *Socrate* alors une proposition comme *Le maître de Platon est humain* équivaut logiquement (sans toutefois être identique) à *Socrate est humain*.

Il est au moins un cas où le concept dénotatif ne peut être ainsi substitué sans dénaturer la proposition et la rendre triviale : en ne prenant en considération que la dénotation, 2, la proposition *2 est 2* serait identique, au niveau logique, à *1 + 1 = 2*[15]. Sans le concept dénotatif, cette dernière serait autant tautologique que la première dès lors qu'on s'abstrairait du texte car "le *est* qui apparaît dans de telles propositions énonce une pure identité" (ici, de 2 avec lui-même) et non pas l'équivalence dénotationnelle des descriptions définies ou des concepts dénotatifs, laquelle est "impliquée mais non assertée" (§64b). C'est l'une des bizarreries de la conception de Russell, qu'une telle proposition affirme une relation d'identité qui porte sur des référents qui ne sont généralement pas incorporés dans la proposition mais contient des concepts dénotatifs, dont l'équivalence est seulement impliquée, et qui confèrent pourtant son sens à la proposition.

Par ailleurs, le choix de Russell de considérer le nom *Socrate* comme un nom propre logique, c'est-à-dire que l'entité *Socrate* est directement présente

---

[15]Ce sont effectivement deux concepts dénotatifs définis singuliers : *la somme de 1 et de 1* et *la somme de 2 et de 0*. Toute expression mathématique exprime ainsi un concept dénotatif. Les concepts dénotatifs définis ont des affinités avec les fonctions : l'imbrication des premiers reflète naturellement la composition des secondes, sans utiliser de variables temporaires superflues.

dans la proposition *Socrate est humain*, a l'attrait immédiat qu'une proposition vraie de *Socrate* est vue, en gros, comme le fait correspondant. Avant même les paradoxes, une telle conception annonce déjà deux apories : le concept dénotatif, dont la dénotation est généralement en dehors de la proposition, laquelle n'est donc pas (seulement ?) un fait, et les propositions fausses, qui n'en sont assurément pas, conduiront à deux épisodes majeurs du démembrement du système[16].

Afin d'assurer un traitement uniforme des objets, Russell aurait pu décider que certains noms propres, voire tous (cf. Frege), ne sont que des descriptions définies déguisées : *Socrate* serait en fait une abréviation pour, par exemple, la description définie *le maître de Platon*. Inversement, on rencontre aussi des noms propres déguisés, lorsque par exemple un philosophe académique emploie la description notoire comme circonlocution afin d'éviter les répétitions lexicales (*l'Athénien*). Une expression figée telle que *sa socratité* serait très proche du nom propre. La question se pose également avec *humain* car *l'humanité* ressemble beaucoup à un nom propre déguisé qu'il serait judicieux d'écrire plutôt *l'Humanité*, qui désignerait rigidement[17] la collectivité des individus réputés humains sur Terre.

Russell voit dans le concept dénotatif un autre critère d'aptitude à la prédication, lui aussi insuffisant : "la propriété la plus [caractéristique] des

---

[16]Son article séminal "On Denoting" [14] éliminera les concepts dénotatifs et le problème des propositions fausses aboutira à l'impasse de sa "Multiple Relation Theory of Judgment". L'assimilation des propositions vraies aux faits est une interprétation répandue mais mal attestée. Dans les *Principles*, Russell utilise souvent le mot *fait* mais il ne l'associe jamais à celui de *proposition* sauf en §430, pour les "prémisses ultimes", et donc sans doute pas les propositions complexes, si bien que certaines objections ultérieures sont sans objet. Et les propositions des *Principles* ne se réduisent pas aux faits de la réalité : la proposition serait plutôt la disjonction des faits possibles qu'elle décrit. Néanmoins l'aporie subsiste pour les propositions fausses comme $2 + 2 = 5$. Une conception du fait réaliste avec son association *directe* à la proposition vraie correspondante, ne permet pas d'analyser un fait faux. Parmi la multitude de solutions proposées par la suite, aucune ne semble apte à compléter adéquatement et de façon naturelle *ce* système. Cette aporie fait suite à celle de l'unité de la proposition "considérée" où l'unité de l'assemblage implique ipso facto sa vérité factuelle.

[17]Une expression X n'est pas rigide si X *aurait pu ne pas être X* a un sens et est vrai. Par exemple, *Le Président aurait pu ne pas être le président* (s'il n'avait pas été élu) est vrai, du moins en France, en l'espère. *Russell aurait pu ne pas être Russell* (le stéréotype qu'on connaît) n'est possible que dans une interprétation très particulière. Il en va de même pour *L'humanité aurait pu ne pas être l'humanité* (telle qu'on la connaît). Ces deux cas ne remettent pas en question la référence de façon essentielle : certes, peut-être, dans ce monde imaginé, *Russell* serait constitué d'autres atomes et *l'humanité* inclurait d'autres individus, mais pour être dans un cas équivalent au premier exemple, il aurait fallu que *Russell aurait pu être Frege* ou *L'humanité aurait pu être les chats* eussent été envisageables. Le caractère de nom propre de *l'humanité* est aussi attesté par la relative difficulté à dire *l'humanité de France* comme on dirait *les humains de France* ou *la faune de France*.

prédicats est leur lien avec la dénotation" (§48b, 46b) c'est-à-dire l'existence d'un concept dénotatif "associé". En effet, un prédicat peut être d'extension vide. Certes, Russell associe à de tels prédicats des concepts dénotatifs, tel *"rien* [qui] est un concept dénotatif qui ne dénote rien" (§73e), de dénotation vraiment vide puisque "il n'y a pas de chose qui soit une classe vide, quoiqu'il y ait des prédicats vides" (§69), mais cela ne fait que transformer le critère inefficient, au mieux, en tautologie, car il faudrait statuer que *Socrate* n'engendre pas de concept dénotatif, à la différence de *cercle carré* qui en engendre un, vide.

Cependant, même si *Socrate* ne peut être employé comme prédicat, le prédicat *socratien* existe, lui, bel et bien, alors comme entité donc distincte de *Socrate*. Il admet le nominalisé *socratitude* et dérive un concept dénotatif comme *le socratien*, lequel dénote *Socrate*. On obtient alors le tableau suivant, étonnamment régulier (Figure 3).

nominalisé concept	adjectif prédicat	description c. dénotatif	nom propre entité
défectuosité	*défectueux*	*le défaut*	
humanitude	*humain*	*les humains* *l'humanité*	*l'Humanité*
socratitude	*socratien*	*le socratien*	Socrate Sa Socratité
daltonisme	*daltonien*	*le daltonien*	Dalton
académisme	*académique*	*l'académie*	*l'Académie*
divinitude	*divin*	*la divinité*	Dieu
candeur	*candide*	*le candide*	Candide

Figure 3. du nom propre au concept

Il est à présent bien difficile de voir en *socratien* la version prédicative de *Socrate*. En outre, un argument décisif est que *socratien* est concevable sans que Socrate ait existé, tout comme *humain* n'implique pas absolument l'existence de l'humanité. Or une description définie peut être vide mais certainement pas un nom propre russellien, qui désigne l'entité elle-même. Le cas échéant, le prédicat ne pourrait donc être identifié qu'au concept dénotatif. On peut certes voir *Socrate* comme un concept dénotatif dérivé de *socratien* pointant rigidement sur Socrate mais ce n'est pas le choix de Russell.

Si *humain* ne correspond à rien du côté de *Socrate* pris comme analogue de *humanitude* et, en particulier, pas à *socratien*, il reste à examiner si *socra-

tien correspond à quelque chose du côté de *humanitude*. Puisqu'un nom propre désigne une chose et qu'un concept n'est qu'une chose particulière, "les concepts peuvent être des objets et avoir des noms propres" (§476) ; ainsi, " "auteur" a pour signification le concept [d'auteur ...] un certain universel [tout comme] "Scott" a pour signification [l'individu] Scott" [15] (p. 216). Cela a une conséquence déroutante : *Socrate* étant désigné, comme *"humain"* = *humanitude*, par un nom propre, si être *socratien*, c'est être identique à l'entité *Socrate*, il serait légitime d'introduire comme analogue un prédicat *humainien*, être identique à l'entité conceptuelle *"humain"*, puis le substantif correspondant, mettons, *humainitude*. On obtient alors le tableau suivant (Figure 4) :

socratitude	socratien	Socrate	S	
humainitude	humainien	"humain" humanitude	humain	les humains

Figure 4. Socrate, analogue de "humain", nom propre du concept

Maintenant, *socratitude* et *humanitude* ne sont clairement plus au même niveau et une utilisation prédicative (*S*) de Socrate serait, le cas échéant, quelque chose de tout différent, bien difficile à imaginer. Il s'avère finalement que *Socrate* semble bien être une chose, et seulement une chose, mais que cette propriété, en tout cas, est encore un "indéfinissable" de plus.

4 Conclusion

Ces remarques illustrent les difficultés et les fondements mal assurés du système des *Principles*, avant même d'en venir à des considérations plus élaborées comme celles liées aux paradoxes, qui concernent le modèle dans la nature même de ses objets et non dans leur utilisation mal régulée. Russell en était certes conscient : dès 1905, dans un article non publié de son vivant ("The Nature of Truth" ; mais cf. aussi §52b), il écrivait par exemple : "les propositions sont des complexes d'une certaine sorte mais je ne sais pas décrire cette sorte de complexité que les propositions ont et que les autres complexes n'ont pas". Sa construction ne permettait pas davantage d'expliquer les propriétés des propositions, dont leur vérité ou (plus encore) leur fausseté. La construction de la molécule sémantique pose des problèmes spécifiques, différant fondamentalement de ceux de la chimie sur deux points au moins : l'absence de substrat géométrique et le caractère orienté de l'assemblage[18].

Dans les *Principia* [17], tous les objets problématiques sont éliminés : concepts dénotatifs, classes, propositions[19]. La méthode, inaugurée avec les descriptions définies dans "On Denoting" [14], est radicale : si ontologiser un objet X (en fait, tous les objets d'un certain type) est gênant, on considère X comme une simple commodité de langage et on édicte des règles, généralement dépendantes du contexte F, qui transforment toute expression $F(X)$ en une expression F' équivalente ne faisant pas intervenir X. En moins d'une décennie d'atomisation logique, ce traitement substitutionnel aura éliminé les objets structurés de la réalité russellienne pour ne garder qu'une nuée d'atomes, que désignent des variables et les noms propres qui ont survécu.

Le système des *Principles of Mathematics* est loin d'une théorie combinatoire adéquate des connexions sémantiques, et des systèmes formels se sont depuis imposés pour explorer et décrire le monde mathématisable. Le système russellien pose néanmoins des questions pertinentes pour l'analyse du langage naturel (dont il est resté proche) et la conception de systèmes cognitifs. Il s'agit alors non plus de décrire la réalité elle-même mais de décrire comment nous pouvons la comprendre.

BIBLIOGRAPHY

[1] Gac, P.: Le complexe dénotant défini de Russell. Noesis **5.2.** (2003)
[2] Geach, P.T.: Subject and Predicate. Mind **59:236**, 461-482 (1950)
[3] Geach, P.T.: *Logic Matters*. University of California Press (1972)
[4] Gödel, K.: Russell's Mathematical Logic. In: Schilpp, *The Philosophy of Bertrand Russell* (1951)
[5] Hochberg, H.: The Wiener-Kuratowski Procedure and the Analysis of Order. Analysis **41:4** (1981)
[6] Hochberg, H.: Russell's early analysis of relational predication and the asymmetry of the predication relation. Philosophia **17:4**, 439-459 (1987)
[7] Hylton, P.: *Russell, Idealism, and the Emergence of Analytic Philosophy* (1990)
[8] Hylton, P.: *Proposition, Functions, and Analysis*. Oxford (2005)
[9] Jubien, M.: Propositions and the Objects of Thought. Philosophical Studies **104**, 47-62 (2001)
[10] Linsky, L.: Terms and Propositions in Russell's Principles of Mathematics. Journal of the History of Philosophy **26:4**, 621-642 (1988)

---

[18] L'assemblage est orienté, non seulement au niveau des connexions, corollairement à l'opposition entre sujet et prédicat, mais encore au niveau même de la structure, corrélativement à la quantification au sens large.

[19] Tout au moins les propositions quantifiées [4] ; diverses nuances sont possibles dans l'interprétation. En effet, que des énoncés soient traités comme des symboles incomplets ne prouve pas qu'il ne leur correspond pas de propositions mais, simplement, et c'est l'essentiel pour ménager une échappatoire aux antinomies, n'oblige plus à admettre d'emblée leur existence. Mais, même s'ils existent, ces objets ne font plus vraiment partie du système.

[11] Linsky, L.: The Unity of the Proposition. Journal of the History of Philosophy **30:2** (1992)
[12] Russell, B.: *Principles of Mathematics*. Cambridge University Press (1903)
[13] Russell, B.: The Nature of Truth (manuscrit non publié, 1905). In Russell, B.: The *Collected Papers of Bertrand Russell* **vol. 4**. Foundations of Logic (1903-1905). (1994)
[14] Russell, B.: On Denoting. Mind **14:56**, 479-493 (1905)
[15] Russell, B.: Knowledge by Acquaintance and Knowledge by Description. In: Russell, *Mysticism and Logic*. Routledge (1911)
[16] Strawson, P.F.: Subject and Predicate in Logic and Grammar. Ashgate (1974, 2004)
[17] Whitehead A.N. et Russell B.: *Principia Mathematica*. Cambridge University Press, Cambridge **vol. 1** (1910)

Philippe GAC
Laboratoire SPHERE (équipe REHSEIS), CNRS, UMR 7219
Université Paris Diderot, Paris, France

Bloody-Minded Revolutions (An Essay Inspired by the Wittgensteinian *Jaha!*)

KATARZYNA GAN-KRZYWOSZYŃSKA & PIOTR LEŚNIEWSKI

— *Man erwarte nur nicht, daß ich mit Achtung von Leute spreche, welche die Philosophie in Verachtung gebracht haben.*
[32, 49]

— *Comment opposer alors l'idée de Rien à celle de Tout?*
[5, 322]

— *Philosophers have perennially argued for particular pictures against their rivals.*
[13, 339]

Introduction

Over one hundred years after the *mathematical turn* in logic, the so-called *practical turn in logic* was declared in [14, 15]. These authors have also written: *It is difficult to overestimate the significance of the mathematical turn in logic*. There are many issues relating to the revolutionary aspects of the changes taking place in contemporary formal logic. For instance: is *mathematosis* just a by-product of the mathematical turn? [29, 127-129] If one appreciates these changes, should one elucidate their outsets?[1] This would require a broad philosophical perspective, for revolutions in logic have to be sketched. In our opinion, the following *recommendation* should be taken seriously in the case of present-day logic: *Back to philosophy*.[2] In the context of the next turn in logic, one should refer to Franz Brentano's theory of the four phases of philosophy [Brentano 1968]. Let us recall that investigations during the second phase (i.e. the first phase of decline: *das erste Stadium des Verfalles*) are determined by practical motives [24]. This begs a "tough" question: is it not the case that the practical turn in logic foreshadows the first stage of the decline in the history of modern logic?

---

[1] It seems that a dynamics of "logic" is a natural part of the universal logic project. For this project, see for example [6].

[2] [40], 121. Let us emphasize the following phrase: *Discussions of 'normative' versus 'descriptive' views of logic have becoming dogmatic, predictable, and boring.* This is a delightful conviction.

From a practical point of view, we place extreme emphasis on following Russell's claim: *Logic used to be thought to teach us how to draw inferences; now, it teaches us rather how not to draw inferences* [31, 151].

We begin with a short (but important) remark devoted to the astounding phenomenon of revolutionary shifts. The stupendous conversion to instrumentalism is briefly presented in Part 2. This is perhaps one of the most instructive and momentously radical changes in the history of modern logic. We move on to associations between logic and metaphysics (Part 3). A curious historical example of such associations is presented – a stimulating lecture delivered by Kazimierz Twardowski (the Founding Father of the Lviv-Warsaw School) on November 18th, 1895 (Part 4).[3] According to this lecture, metaphysics is aimed at achieving one true philosophical system. Jules Vuillemin's idea of an *a priori* classification of alternative philosophical systems and his assumption that every form of predication becomes an ontological principle in a genuine philosophical system are also discussed (Part 5). Finally, the concept of the solvability of ontological questions is introduced, and a consideration of metaphysical theorems strictly connected with classical logic is provided, which is simultaneously an uncomplicated exercise in propositional calculus (Part 6). In place of a conclusion, our paper ends with some remarks on bloody-minded revolutions in art.

However, first a note – and then a question – in [43] was our real inspiration. Among other things, Wittgenstein wrote on September 3, 1914: "Do any of the forms exist at all that Russell and I were always talking about? (Russell would say: 'yes! that's self-evident.' *Ha!*)" [43, 3/3e]. Indeed, what happens if the Russellian answer is rejected?

In talking about bloody revolutions, we would like to refer to a lionized phrase from the famous speech *Message to the Grass Roots* delivered by Malcolm X on November 10th, 1963: *Revolution is bloody, revolution is hostile, revolution knows no compromise, revolution overturns and destroys everything that gets in its way*. Two fundamental questions in theories of social revolutions thus need to be mentioned: (A) How to explain the explosion of involvement, participation, activism, and disobedience which is observed at the beginning of each revolution?; and (B) Why are the results of a revolution something entirely different from the dreams of the revolutionaries?[4] Question (B) could be reformulated as: (C) Why do bloody revolutions become usually bloody-minded ones? It is said that someone is bloody-minded if he/she deliberately making things difficult for other people instead of being helpful. Unquestionably, the relative directive is: Stop being bloody-minded! Don't make things awkward.

---

[3] Needless to say, Łukasiewicz was a disciple of Twardowski.
[4] See, for example, [35], 303; and [25]. Cf. [26]. For revolutions in science, see [18].

2. A send-off mythology (1): Łukasiewicz

Without any doubt, the Polish School of Logic was a mathematical one. But it was a philosophical school, as well [34, 379].[5] Jan Łukasiewicz stated explicitly in [20] that logic is a science about logical values – therefore logic has its very own subject of investigation. No other science is concerned with this subject. Moreover, logic is not a science about sentences, because such study is the domain of grammar. Consequently, logic is not a science about contents expressed by means of sentences. Logic does not deal with propositions or beliefs – this pertains to psychology. Logic is not an ontology, since the latter is related to inquires into "objects at all". Logical values are two (different) objects that are denoted by sentences. There are as many different names for a sole, singular logical value (i.e. the value called "Truth") as we have true sentences. Ontologically, Truth corresponds with being/existence, and Falsity with non-being/non-existence, respectively. Many years later, Łukasiewicz himself called this point of view a *mythology*, writing in 1952:

> Logic is not a science of the laws of thought or any other real object; it is, in my opinion, <u>only an instrument which enables us to draw asserted conclusions from asserted premises</u>. The classical theory of deduction which is verified by a two-valued matrix is the oldest and simplest logical system, and therefore the best known and widely used. But for some purposes, for instance in modal logic, an n-valued system, $n > 2$, might be more suitable and useful. The more useful and richer a logical system is, the more valuable is. [21, 333, emphasis ours][6]

The big question that arises is why Łukasiewicz changed his view on logic? There are a lot of tiny treasures in Łukasiewicz's papers, a collection of which has been published [22]. For example, in a letter to J. M. Bocheński from Dublin (on September 4th, 1946), Łukasiewicz wrote that he does not

---

[5]Cf. [33]. We do not discuss here such historical (and even "piquant") names as the *Warsaw Circle* and the *Berkeley-Warsaw School*. The first name has been used by Alfred Tarski in a letter to Otto Neurath (September 7th, 1936); see [36]. For the second name, see [45]. Let us recall the old and intriguing remark from [2], 38: *Leaving historical disputes aside, it is a matter of urgency that a doctrine should be developed about the various kinds of good reasons for which we 'call different things by the same name'. This is an absorbing question, but habitually neglected, so far as I know, by philologists as well as by philosophers. Lying in the no man's land between them, it falls between two schools, to develop such a doctrine fully would be very complicated and perhaps tedious: but also very useful in many ways. It demands the study of actual languages, not ideal ones. That the Polish semanticists have discussed such questions I neither know <u>nor believe</u>* [emphasis ours].

[6]It should be noted that the paper 'On the Intuitionistic Theory of Deduction' was published in *Konikl. Nederl. Akademie van Wetenshappen, Proceedings, Series A*, No. 3, 202-212. It is also worth emphasizing that the name *Łukasiewicz's logics* is an ambiguous phrasing even in the Polish tradition. See, for instance, [44], 68. For a beautiful and perfect example of research onto many-valued logics, see [17].

like, among other things, the algebra of logic, since he does not accept empty classes [22, 519].

It is our conviction that if we want to understand the rapid change called *the mathematical turn in logic*, we should study the main sources of this revolution.

3. In the terrible grip of boring beauty: metaphysics and logic

Let us begin with a gorgeous phrase by Wang (1974): *There is more philosophical value in placing things in their right perspective than in solving specific problems.* [42, x, emphasis ours][7]. The term *philosophical value* is our point of reference. Why? First, because of the famous remark by Frege: *Like ethic, logic can also be called a normative science* [39, 180]. A contrast with Carnap's conviction should be accentuated here: *In logic, there are no morals* [11, 52]. Maybe Carnap was right; however, this does not mean that some axiological questions are inadmissible in the historical contexts of modern logic. Following Pihlström, we repeat that philosophy is often presented as a game of argumentation. At the same time, we find his model of an integration of ethics and metaphysics thought provoking [27, 667-669].

Let us examine then briefly some associations between logic and metaphysics, since we are going to suggest some sort of integration between them at the end of this paper.[8] We can start with an excellent note from Wittgenstein (1979): *Philosophy consists of logic and metaphysics: logic is its basis.*[9] We are not investigating here a relation: "a philosophical discipline x is a basis for a philosophical discipline y". We are merely putting forward the following thesis (without any further justification here of course):

(*) Philosophy consists of metaphysics and logic: metaphysics is its basis.[10]

Since we hold erotetic logic in high esteem, we shall adumbrate another technique. Five questions are introduced: (Q_1) Is logic (as the science of reasoning) a philosophical discipline? (Q_2) Has metaphysics (as a philosophical discipline) its own subject-matter? (Q_3) Is the construction of a single,

---

[7]In *Philosophy of Logic* Susan Haack writes: *It is often enough the case in philosophy that asking the right question is half the battle.* At the workshop *Logic in Question* (Paris, May 2-3, 2011) we have put informally the question: *What about the other half?* See [16], 227.

[8]For an introduction to metaphysics, see, for example, [19], and/or *Metametaphysics*, 2009.

[9]Wittgenstein 1979, 106. For *the Wittgensteinian Jaha!* see ibid, 3/3e.

[10]For other approach see [13].

solitary philosophical system the final aim of metaphysics? (Q_4) Has there been some sort of (historical, theoretical) progress within the framework of metaphysical issues? (Q_5) Should scientific methods (for example deduction, statistics) be applied within the framework of metaphysical studies? Let us take these questions as simple yes-no questions. Therefore, each of them has only two answers: (1) a *positive* answer and (2) a *negative* answer. Hence, we arrive at 32 different standpoints. The first standpoint consists of positive answers to these questions exclusively and so on.

In relation to the first question, one should notice that the so-called logic (1) is treated as *that branch of philosophy in which we discuss the nature and criteria of reasoning* [12, 65, emphasis ours]. We take other paradigmatic illustrations of such standpoints from Kazimierz Twardowski and Jules Vuillemin.

4. A farewell to post-Kantian speculations: Twardowski

Pouivet has written: *S'agissant de la philosophie contemporaine, il est malaisé de fixer la date de naissance (...) Pour ma part, la leçon inaugurale de Kazimierz Twardowski à l'Université de Lvov, en 1895, est le moment décisif* [28, 11]. So let us take a glimpse at some of the theses included in a lecture delivered by Twardowski on November 18th, 1895. Philosophy is a collection (not a system) of philosophical disciplines (logic, psychology, metaphysics, ethics, and aesthetics, among others). However, the final aim within the framework of metaphysics consists of the construction of one, single philosophical system.

There is the famous fourth Brentano's *Habilitationsthese* [1, 437]. Let us begin with the short historical sketch of philosophy in Twardowski's lecture:

> Since there are no two equal souls, everyone has started with different assumptions and came to different conclusions. This is the cause of multiplicity of rivaling systems and philosophers' eternal quarrel, total lack of agreement had generated general mistrust and then disdain and disregard. Philosophy created one apriorical system after another, philosophers strained and generated new ideas, while natural sciences slowly but continuously progressed by means of quiet, systematic work based on facts. Copernicus, Galileo, Newton, Darwin, Robert Mayer, Kirchoff, Helmholtz, Fechner, Maxwell and Hertz show the way how reach general principles on the base of facts and induction. These principles would serve as a foundation for the most important deductions that could have been verified by experience. For philosophy it was the end of line. After the failure of post-Kantian speculation when intellectual vacuum arose, some superficial minds tried to fill this gap with gospel of materialism, and then philosophy awakened. There Herbart and Trendelenburg, Lotze and Brentano had risen and called to repair the pitiful state. They had rejected pseudo-aprioristic theses which were in fact ar-

bitrary; they indicated an analysis of the brute facts and phenomena themselves, they put the method of natural sciences forward as a model and then a new era in philosophy, especially in metaphysics has begun [38, 232-233, authors' translation, emphasis ours].

The final remarks in Twardowski's lecture begin with following phrase:

I tried to prove that between philosophical and natural sciences there is no such a great striking discrepancy; on the contrary, one of the branches of philosophy, namely metaphysics considers issues concerning the rest of philosophical disciplines as well as natural sciences. I also tried to show that the methods of philosophical sciences including metaphysics are exactly the same as are normally used in the research within natural sciences. In the present condition of philosophy, in consequence, philosophers should be no longer classified and labeled [38, 235, emphasis ours]

Twardowski has said that the time for the construction of the one true philosophical system has not yet come. Moreover, we do not have such a system as yet, and perhaps we shall never achieve it. He explained:

Mechanics managed to reach the perfect system, because its subject is precisely defined, its limits are determined and its analysis refer to rather simple and not complicated phenomena. However, metaphysics should embrace the whole universe, its past and future, everything what is infinitely small and infinitely large! Will metaphysics ever come to embrace all these particular information, which will build the complete system of knowledge? I do not know the future, but I know that today we are very far from this complete system but not as far as Indic philosophers and Ionian philosophers [38, 234].

He also formulated a brief schedule:

Let us not examine the future, let us not formulate speculations, instead let us do what we should do. Since we admitted that we are not yet able to construct the system of philosophy, let's collect information that would be useful in the future. We are not missing issues that we could consider, so let us analyze them one after another, and so we will deliver more and more foundations to humanity, more and more data, that could be useful in better understanding of the universe [38, 234].

5. A send-off mythology (2): Vuillemin

Regarding a relation between society and philosophy, Vuillemin considered that not every civilization is able to disentangle itself from a certain bondage to law, custom, and social utility. Hence, [f]*ree philosophy, therefore, as opposed to Church- or state-organized ideology, must have arisen from a revolution in the use of signs by means of which, for every civilization, language represents the sensible world* [41, 96, emphasis ours].

According to Twardowski, there is one true philosophical system, but we are not yet able to construct such a system. A different approach is proposed by Vuillemin – the *a priori* classification of alternative philosophical systems (of ontologies – in other words). This classification describes possible classes of dogmatic systems such as: realism, conceptualism, nominalism of things, nominalism of events and systems of examination: intuitionism and skepticism. Four fundamental assumptions characterize Vuillemin's project. Firstly, he denies that language shapes perception, since perception precedes language. Secondly, the so-called *Weltanschauung* (a world perspective) is not a philosophy. Thirdly, a strong continuity between philosophy and common sense is explicitly rejected. Finally, Vuillemin does not argue for a unique scheme of philosophical truth. He rather wants to consider what all the possibilities of truth are. It is worth emphasizing that Vuillemin outlines a radical break between myth (mythological thinking), which is connected with the problem of *Weltanschauung*, and free philosophy. According to Vuillemin, free philosophy definitely begins with axiomatization: *philosophy and science have common origin in the discovery of antinomies and in the development of axiomatics, I deny that strong continuity between philosophy and common sense which is advocated by many philosophers of natural language* [41, viii].

The principles of philosophical systems are directly connected with linguistic categories and with the primitive elementary sentences by which they are expressed. According to Vuillemin, natural language as an open, dynamic and universal code of communication contains five exclusive classes of elementary sentences: pure predication, substantial and accidental predication, circumstantial predication, judgments of methods, and judgments of appearances.[11] Every form of predication becomes an ontological principle of a genuine philosophical system (ontology). A systematic ontology therefore requires the following conditions: (a) there must be given a minimal set of *indefinable concepts* and *indemonstrable principles* from which all the elements of the world can be derived, (b) such a derivation should proceed according to *legitimated rules*, and (c) rival ontologies should be explained away as mere appearances.

The development of philosophy is correlated with the development of the axiomatic method. Philosophy and axiomatics give signs a new interpretation. Vuillemin wrote: *Axiomatic method and philosophy gave this revolution its expression in both the domain of the particular sciences and the domain of rationality in general. The revolution resulted in man's struggle to disentangle reality from appearances.* [41, 96, emphasize ours]. Let us shortly consider the opposition between myth and science. In order to

---

[11] For individuals, in a given social structure or/and in a given social context, to be able to communicate their perceptual experiences, two conditions must be fulfilled. Their system or code of communication must be universal and open.

be possible, science requires: (I) a complete, total *revolution* in the use of linguistic signs; (II) a theoretical determination; and (III) a theory of truth.

There are five exclusive classes of predication, since the determination of the singular term occurs in respect to either: (1) the general term, (2) the singular term itself, (3) the material chain of the sentence, (4) its syntactical unit, or (5) its semantic unit. The correlated classes of elementary sentences include: pure predication, substantial and accidental predications, circumstantial predication, judgment of method and judgment of appearances.

The five kinds of identification (1)–(5) divide into two major series, distinguished in terms of truth, i.e. a dogmatic series (which consists of: pure predication, substantial predication, accidental predication, circumstantial predication), and subjective series. The word *dogmatic* means that such sentences refer and identify without relying on *the subjectivity* of the speaker. The formal principles of identification relevant to the predication of a dogmatic series depend on objective structures, institutions or pragmatic decisions, that is, the code of the language, the conventions of the group, or the relation between the speaker and his/her message. The subjective series consists of judgments of method and judgments of appearance. Both forms of predication comprehend the speaker's possible subjective contributions to the building of the sentence [41, 47].

Thus the Vuillemin's approach includes a great idea: a coexistence of different ontologies (classes of ontologies). This needs further exploration. It is beyond all doubt.

6. A farewell to the history of philosophy: on the solvability of ontological problems

However, let us relinquish the intricacies of contexts from the history of philosophy just for a moment. For instance, let's rid ourselves of the Quinean slogan *no entity without identity*. Moreover, let's be logical within the framework of metaphysics! A theory of existence that is tightly connected with classical (i.e. two-valued and at the same time extensional) calculus will be proposed here. The *rationale* of the theory is a consequence of the *truth-table decision method*, and the latter is identified with solvability (of a given problem).[12] Imagine this: you are dealing with existence-questions, and they have answers!

The phrase "to proceed rationally (to behave in rational way)" has a very precise meaning here, to wit "to accept ontological theses". Thus, saying "to proceed perfectly rationally (to behave in a perfectly rational way)" stands for "to accept all ontological theses".

---

[12]The truth-table decision method will be later restricted to *the Sheffer's stroke* (NAND).

Let L_0 be first-order language supplemented with two *erotetic* constants: "?" [question mark] and "{ }" [brackets]. The declarative well-formed formulae (d-wffs) of L_0 are defined in the standard way. By the question of L_0, we understand an expression of the very general form:

(#) ?$\{A_1, \ldots, A_n\}$,

where $n > 1$ and $A_1, \ldots, A_n$ are syntactically distinct sentences of L_0. The concept of direct answer to a question (i.e. the possible and just-sufficient answers to the question) is understood in the standard way. By a direct answer to a question of the form (#), we mean each of the sentences $A_1, \ldots, A_n$. Simple yes-no questions are expressions of the form:

(##) ?$\{A, \neg A\}$.

By *an ontological thesis*, we understand here each propositional tautology (law) of classical sentential logic (classical zeroth/zero order logic) written by means of signs from the alphabet of the particular language L_0.[13] A problem (expressed by a given question) of the form ?$\{A_1, \ldots, A_n\}$ of the language L_0 is *solvable* if and only if at least one sentence in the set $\{A_1, \ldots, A_n\}$ is a propositional tautology of classical sentential logic. Needless to say, classical logic is what we want to retain. In other words, it is what we play for.[14]

From now on, metaphysics (ontology) will be understood here entirely as *a theory of existence* – in the very formal sense of the term *theory*, if you wish. At the moment, the terms *existence*, *being* (the distributive meaning of the word "being"), *Being* (the collective meaning of the word "being"), entity, and so on are synonyms in defiance of the old philosophical tradition. For instance, the concept of a *watered-down* "being" is omitted [4, 167].

Let L_1 be the first-order language *without* identity, but with one extra-logical symbol $ex(x)$. The alphabet of L_1 includes standard propositional connectives, i.e.: $\neg$ (*the classical negation sign*) and $\wedge$ (*the conjunction sign*) at the very least. Consider two extensions of the vocabulary of L_1. First, a constant b (also called *the first fundamental word*) is introduced. The reading of the constant b as *being* is preferred. On the other hand, *Being* (or *the Being*), *entity*, *Entity* (or *the Entity*), *existence*, *Existence* (or *The Existence*), and so on are allowable, as well. There are two simplest sentences of L_1:

(S_1) $ex(b)$ ["Being exists."];

---

[13] For the intuitive concept of propositional tautology see e.g. [30], 163-166.
[14] For example, deviant logics and extended logics are strongly avoided.

(S_2) $\neg ex(b)$ ["Being does not exist."].

Subsequently, the second constant n is added to the vocabulary of L_1. This constant is called *the second fundamental word* and the reading as *nonbeing* is favored. Since we are talking about nonbeing, thereby the two following formulae:

(S_3) $ex(n)$ ["Nonbeing exists."];

(S_4) $\neg ex(n)$ ["Nonbeing does not exist."],

will be called *meontological sentences*. Incidentally, get a load of these four sentences:

(S_5) $ex(b) \land ex(n)$;

(S_6) $\neg ex(b) \land ex(n)$;

(S_7) $ex(b) \land \neg ex(n)$;

(S_8) $\neg ex(b) \land \neg ex(n)$.

All the more, one might expect that two equations are obligatory, viz.:

(E_1) $ex(n) := \neg ex(b)$,
["Nonbeing exists." is equal to "It is not a case that being exists."]

(E_2) $\neg ex(n) := ex(b)$
["It is not a case that nonbeing exists." is equal to "Being exists."].

We do not insist on this. Each sentence of L_1 containing b as the only extra-logical constant could be referred to as *an ontological sentence*. Hence, each sentence of L_1 with n as the unique extra-logical constant could be referred to as *a meontological sentence*.

And now, let L_2 be a first-order language without identity but with one extra-logical symbol $ex(x)$. The alphabet of L_2 includes the unique standard propositional connective, i.e.: $\uparrow$ (*the Sheffer's stroke sign*). The only extra-logical constant in the vocabulary of L_2 is the first fundamental word b. For the sake of simplicity, the succeeding three abbreviations are established:

(D_1) $ex(b) := ex(b)^1$;

(D_2) $ex(b) \uparrow ex(b) := ex(b)^2$.

(D_3) $\alpha \uparrow \alpha := \alpha^2$, for any expression α.

Let us prompt that each propositional tautology of classical sentential logic is an ontological thesis of L_2. By means of the truth-table decision method for Sheffer's stroke, one easily verifies that the following expressions are ontological theses of L_2:

$(T_1)\ ex(b)^1 \uparrow ex(b)^2$;

$(T_2)\ ex(b)^2 \uparrow (ex(b)^2)^2$;

$(T_3)\ ((ex(b)^2)^2 \uparrow ex(b)^2$;

$(T_4)\ ex(b)^1 \uparrow (((ex(b)^2)^2)^2$;

$(T_5)\ (ex(b)^1 \uparrow ((ex(b)^2)^2) \uparrow (ex(b)^2)^2$.

7. Instead of a conclusion: bloody-minded revolutions in art

Following Sabato, we would like to repeat: an artist is a true rebel — a rebel *par excellence* — and that is why an artist is always in trouble during revolutions. In a conversation with Borges, Sabato also claimed that every revolution is moralistic and puritanical. What we would like to point out here is the fact that art created in revolutions is not at all revolutionary; it is academic, always conservative and simply very poor.[15] Therefore, revolutions can be dangerous because they are contrary to free creation; they establish very strict norms and standards as well as a horrendous bureaucracy. The same idea can be also found in Gide's writings.[16] The artist as a real rebel (and not a revolutionary) should not have the conscious purpose of becoming absolutely free. Ergo, art that reflects one's intention, according to Borges, its worth nothing. *Art* just *happens*, as one of Borges' favorite sentences goes. Hence, art cannot be institutionalized if we want to keep it alive. Paradoxically, the worst thing that can ever happen to an artist is to become classic [8, 88]. However, it is also possible that Ingmar Bergman was right when he said:

> "I've a strong impression our world is about to go under. Our political systems are deeply compromised and have no further uses. Our social behaviour patterns — interior and exterior — have proved a fiasco. The tragic thing is, we neither can nor want to, nor have the strength to alter course. It's too late for revolutions, and deep down inside ourselves we no longer even believe in their positive effects. Just around the corner an insect world is waiting for us — and one day it's

---

[15] [3], p. 28-29. They both gave examples of very bad works of art which had been created during revolutions. For instance, during French Revolution instead of Delacroix, an academic painter — David was official artist.

[16] [15], pp. 70-71. In the late thirties, he wrote: *J'écrivais avant d'aller en U.R.S.S. Je crois que la valeur d'un écrivain est liée à la force révolutionnaire qui l'anime, ou plus exactement (...) à sa force d'opposition. (...) le triomphe de la révolution permettra-t-elle à ses artistes d'être portés par le courant? Car la question se pose : qu'adviendra-t-il si l'Etat social transformé enlève à l'artiste tout motif de protestation? Que fera artiste s'il n'a plus à s'élever contre, plus à se laisser porter?*

going to roll in over our ultraindividualized existence. Otherwise I'm a respectable social democrat." [7, 18, emphasis ours].

But logic isn't an art, is it?

Bibliography

[1] Albertazzi, L., 1996, "From Kant to Brentano", in *The School of Franz Brentano*, pp. 423-464.

[2] Austin, J. L., 1961, "The Meaning of a Word", in J. L. Austin, *Philosophical Papers*. J. O. Urmson & G. J. Warnock (eds.), Oxford: Clarendon Press, pp. 23-43.

[3] *Diálogos Jorge Luis Borges Ernesto Sabato*, 2002, Barone O. (ed.), Barcelona: Emecé Editores.

[4] Benzivenga, E., 2002, "Free Logic", in D. Gabbay & F. Guenthner (eds.), *Handbook of Philosophical Logic*, vol. 5, Dordrecht/Boston/London: Kluwer, pp. 148-196.

[5] Bergson, H., 1908, *L'Evolution créatrice*, Paris: Félix Alcan.

[6] Béziau, J.-Y., 2006, "13 Questions about Universal Logic. 13 questions to Jean-Yves Béziau", by Linda Eastwood, *Bulletin of the Section of Logic*, vol. 35, pp. 133-150.

[7] Bjorkman S. & Manns T. & Sima J., 1973, *Bergman on Bergman*, New York: Simon and Schuster.

[8] Borges, J. L. & Ferrari, O., 2005, *En diálogo/I*, México: siglo XXI editores.

[9] Brentano, F., 1968, *Die vier Phasen der Philosophie und ihr augenblicklicher Stand nebst Abhandlungen über Plotinus, Thomas von Aquin, Kant, Schopenhauer und Auguste Comte*, Mit Anmerkungen hergestellt von O. Kraus. Neu eingeleitet von F. Mayer-Hillebrand, Hamburg: Meiner.

[10] van Benthem, J., 2006, "Where is logic going, and should it?", *Topoi*, vol. 25, pp. 117-122.

[11] Carnap, R., 2002, *The Logical Syntax of Language*, Translated by A. Smeaton, Chicago: Open Court.

[12] Curry, H. B., 1951, *Outlines of a Formalist Philosophy of Mathematics*, Amsterdam: North-Holland.

[13] Dummett, M., 1994, *The Logical Basis of Metaphysics*, Cambridge, MA: Harvard University Press.

[14] Gabbay D. M. & Woods J., 2005, "The Practical Turn in Logic", in D. M. Gabbay & F. Guenthner (eds.), *Handbook of Philosophical Logic*, vol. 13, Dordrecht: Springer, pp. 15-122.

[15] Gide A., 2009, *Retour de l'URSS*, Paris: Gallimard.

[16] Haack, S., 2000, *Philosophy of Logic*, Cambridge/London/New York: Cambridge University Press.

[17] Kamide, N., 2005, "A Cut-free System for 16-Valued Reasoning", *Bulletin of the Section of Logic*, vol. 34, n° 4, pp. 213-225.
[18] Kuhn, T. S., 1962, *The Structure of Scientific Revolutions*, Chicago/London: The University of Chicago Press.
[19] Lopston, P., 2010, *Reality: Fundamental Topics in Metaphysics*, Ottawa: University of Ottawa Press.
[20] Łukasiewicz, J., 1920, "Logika dwuwartościowa" ["Two-valued Logic"], *Przegląd Filozoficzny*, vol. 23, pp. 189-205. See also in (Łukasiewicz 1998, pp. 110-125). [Polish]
[21] Łukasiewicz, J., 1970, "On the Intuitionistic Theory of Deduction", in *J. Łukasiewicz Selected Works*, L. Borkowski (ed.), Amsterdam-London: North-Holland, Warszawa: PWN-Polish Scientific Publishers.
[22] Łukasiewicz, J., 1998, *Logika i metafizyka. Miscellanea pod redakcją J. J. Jadackiego*, [*Logic and Metaphysics. Miscellanea edited by J. J. Jadacki*]. Warszawa: Wydział Filozofii i Socjologii Uniwersytetu Warszawskiego. [Polish]
[23] *Metametaphysics: New Essays on the Foundations of Ontology*, 2009, D. J. Chalmers, D. Manley, R. Wasserman (eds.), Oxford: Clarendon Press.
[24] Mezei B. M. & Smith, B., 1998, *The Four Phases of Philosophy*, Amsterdam-Atlanta, GA: Rodopi.
[25] Nowak, L. & Paprzycka & K., Paprzycki, M., 1993, "On Multilinearity of Socialism", in L. Nowak & M. Paprzycki (eds.), *Social System, Rationality and Revolution, Poznań Studies in the Philosophy of Sciences and the Humanities*, vol. 33, pp. 355-370.
[26] Ost. D., 2005, *The Defeat of Solidarity: Anger and Politics in Postcommunist Europe*, New York: Cornell University Press.
[27] Pihlström, S., 2009, "Ethical Unthinkabilities and Philosophial Seriousness", *Metaphilosophy*. vol. 40, n° 5, pp. 656-669.
[28] Pouivet, R., 2008, *Philosophie contemporaine*, Paris: Presses universitaire de France.
[29] Quine, W. V., 1987, *Quiddities. An Intermittently Philosophical Dictionary*, Cambridge, MA: The Belknap Press.
[30] Rasiowa, H. & Sikorski, R., 1963, *The Mathematics of Metamathematics*, Warsaw: National Scientific Publishers.
[31] Russell, B., 1985, *The ABC of Relativity*, London: Unwin Hyman.
[32] Schopenhauer, A., 1890, "Über die vierfache Wurzel des Satzes vom zureichenden Grunde. Eine philosophische Abhandlung", in *Arthur Schopenhauers Sämmtliche Werke*, Erster Band. Stuttgart: J. G. Cotta, pp. 33-191.
[33] Sundholm, G., 2003, "Tarski and Leśniewski on Languages with Meaning versus Languages without Use. A 60th Birthday Provocation for Jan Woleński.", in J. Hintikka, T. Czarnecki, K. Kijania-Placek, T. Placek and

A. Rojszczak (eds.), *Philosophy and Logic. In search of Polish Tradition.* Dordrecht: Kluwer, pp. 109-128.

[34] Suszko, R., 1977, "The Fregean Axiom and Polish Mathematical Logic in the 1920's", *Studia Logica*, vol. 36, 4, pp. 377-380.

[35] Sztompka, P., 1999, *The Sociology of Social Change*, Oxford: Blackwell.

[36] Tarski, A., 1992, "Drei Briefe an Otto Neurath", R. Haller (ed.), with an English translation by J. Tarski. *Grazer Philosophische Studien.* vol. 43, pp.1-32.

[37] *The School of Franz Brentano*, 1996, L. Albertazzi, M. Libardo, R. Poli. (eds.), Dordrecht: Kluwer.

[38] Twardowski, K., 1994, "Wykład wstępny w Uniwersytecie Lwowskim" (z 15. listopada 1895r.), [An Inaugural Lecture at the Lvov University (on November 18th, 1895)], *Principia*, vol. 8-9, pp. 225-236. [Polish]

[39] Vaidya, A. J., 2006, "The Metaphysical Foundation of Logic", *Journal of Philosophical Logic*, vol. 35, pp. 179-182.

[40] Van Benthem, J., 2006, "Where is Logic Going,and Should It?", *Topoi*, Vol. 25, pp. 117-122

[41] Vuillemin, J., 1986, *What are philosophical systems?*, Cambridge: Cambridge University Press.

[42] Wang, H., 1974, *From Mathematics to Philosophy*, London: Routledge & Kegan Paul.

[43] Wittgenstein, L., 1979, *Notebooks 1914-1916*, G. von Wright & G. E. M. Anscombe (eds.), with an English translation by G. E. M. Anscombe, Oxford: Basil Blackwell.

[44] Wójcicki, R., 1984, *Lectures on Propositional Calculi*, Wroclaw: Ossolineum/The Publishing House of the Polish Academy of Sciences.

[45] Wójcicki, R., 2003, "Polish Logic of the Postwar Period", [in collaboration with Jan Zygmunt] Retrieved from:
http://www.ifispan.waw.pl/studialogica/PL.Logic.html#note4

Katarzyna GAN-KRYWOSZYNSKA and Piotr LESNIEWSKI
Department of Logic and Methodology of Sciences, Institute of Philosophy
Adam Mickiewicz University, Poznań, Poland

Formal Logic and Ancient Mathematical Reasoning

PIERLUIGI GRAZIANI AND MAURIZIO COLUCCI[1]

1 Introduction

The present paper intends to address the question whether formal logic constitutes a valuable instrument for analysing ancient mathematics. Starting from the mid-seventies[2], the question has been the object of a new wave of interest. Nonetheless, in its most general form, it can be traced back to syllogistic renditions of Euclids *E*lements and to the more recent formal renditions of syllogistic logic[3].

Single case studies that we hold relevant for our purposes[4] have focused on: 1) the ancient method of analysis - synthesis[5]; 2) the role played by visual reasoning in ancient science[6]; 3) the question whether and why Greek ancient mathematical proofs have general scope[7]; 4) ancient reflections on divisibility and infinite divisibility[8].

In what follows, we will restrict our attention to the interpretation of Euclid's Postulates.[9]

---

[1] We are very grateful to Adriano Angelucci, Massimiliano Badino, Claudio Calosi, Giuliano Torrengo, Vincenzo Fano and anonymous referees for their comments on different drafts of this paper.

[2] Hintikka and Remes [1974; 1976]; Mueller [1981]; Mäenpää and von Plato [1990]; Mäenpää [1993; 1997]; Mumma [2006]; Graziani [2007; 2014]; Miller [2008]; Mumma, Avigad and Dean [2009]; Beeson [2009; Ts]

[3] With regard to the syllogistic renditions of Euclids Elements see: Mugnai [2010]; Bertato [204]. With regard to the formal renditions of syllogistic logic see, for example, Łukasiewicz [1957]; Corcoran [1974]; Thom [1996]; Ridolfi [1999] and their detailed bibliographies.

[4] The works listed in Bibliography contain rich and updated references. Readers should refer to them for classic works on the subject.

[5] See, in particular, Hintikka and Remes [1974]; Hintikka and Remes [1976]; Mäenpää [1993]; Sidoli and Saito [2012]

[6] See, in particular, Luengo [1995]; Miller [2001]; Miller [2006]; Mumma [2006; 2010; 2012]; Miller [2008].

[7] See Hintikka and Remes [1974]; Mueller [1981]; Mäenpää [1993]; Acerbi [2011].

[8] See Calosi and Fano [2015].

[9] See section 2 for the reasons of this choice.

The case studies mentioned above are all naturally related to one another and therefore suggest the possibility of providing a unified account of ancient mathematics and its methods of discovery and justification. Such an achievement, as we will argue, can remarkably profit from an extensive and systematic use of formal methods of analysis.

By the expression *'Formal History of Sciences'* (*FHS*) we will refer, throughout the paper, to any analysis of scientific problems in the History of Sciences which employs formal techniques. *FHS*, on our view, not only represents an important application of formal logic, but it can also valuably contribute to the advancement of this discipline[10]. Its underlying motivation is the belief that adequate formal translations can reveal the logical structure of principles and arguments which, due to the known opacity of natural language and frequent lack of textual evidence, would not otherwise be immediately visible. In particular, it proves very useful in detecting logical mistakes, implicit premises and valid arguments within the informal theory under investigation. It follows, we believe, that *FHS* should be counted among the instruments that we normally rely on in our historical analyses of sciences. Despite the fact that several existing studies could plausibly be considered instances of *FHS*, this approach has not yet reached the status of a fully established methodology. The current state of the art looks rather like a disorganized aggregate of independent results. Our impression, in particular, is that while some formal results are provably useful[11] in reconstructing scientific thought, some others run the risk of being misleading[12].

As a consequence of this situation, we felt the need to work, with respect to the case of ancient mathematics[13], on a more rigorous characterization of *FHS*. In what follows we will speak accordingly of *Formal History of Mathematics* (*FHM*) in order to refer to this kind of endeavor.

Nathaniel Miller [2008, 12] has rightly observed that:

> In a formal system, we have a clearly defined notion of whether or not a proposed inference rule is valid —it is valid if it always gives a true conclusion when provided with true hypotheses. Euclid's proofs, however, are not part of a formal system in the modern sense, so we can't apply this test to them. This, by itself, doesn't mean that his methods are incorrect—only that they are informal. There are certainly informal proofs that mathematicians accept as being correct; in fact, practicing mathematicians almost never give proofs of their results in a formal system. Rather, they accept the idea that a formal system for doing mathematics exists, and if pressed might claim that their proofs could be translated into such a formal system.

---

[10]See, in particular, Mäenpää [1993; 1997; 1998]; von Plato [1995; 1998]; Graziani [2007; 2014]; Beeson [2009; Ts; 2015].

[11]See Graziani [2007; 2014].

[12]With regard to the history of ancient mathematics, for instance, this is the case of Robinson [1936] and of some ideas contained in Hintikka and Remes [1974; 1976] and Mueller [1981].

[13]A more detailed analysis will be presented in Graziani [2016a; 2016b].

Although translating an informal proof into a formal system often turns out to be a very difficult task, we agree with Miller's observation and feel therefore inclined to take his hypothesis as fundamentally correct[14]. Nonetheless, we shall restate it in the following, more general, form:

Formality Hypothesis:
An informal reasoning method is correct if and only if it is possible to lay out a formal system with the property that informal reasoning using the informal methods can always be translated into equivalent correct formal reasoning in the formal system.

This tentative principle, as we would like to suggest, constitutes an interesting criterion to evaluate informal reasoning methods[15]. Its provisional status is due the fact that, at present, the hard question as to what exactly constitutes a 'formal equivalence' remains open[16]. This is what makes our proposal a 'formal hypothesis' rather than a 'formal conjecture'.[17]
As it stands, the hypothesis can be elaborated further into some general criteria for the development of *FHM*. A proposal in this direction has been advanced in Graziani [2016b], which argues in favor of five further fundamental criteria of adequacy for formalization, namely the *Principle of Expressivity, Principle of Completion, Principle of Reliability, Principle of Ambitiousness*, and the *Principle of Transparency*.[18] The general idea is still that of providing an answer to our initial question, concerning the usefulness of formal logic for the analysis of ancient mathematics. We believe that, if the use of formal logic is informed by principles establishing clearcut formalization criteria, this question can be answered in the affirmative. We will focus, in what follows, on the *Principle of Expressivity* only[19]. Although a rigorous assessment of *FHM* (*FHS*) would require an analysis of all the criteria (and of the relations holding amongst them), defending the

---

[14] See also Feferman [2012].

[15] To put it in Miller's words "it gives us a basis for evaluating informal methods of proof".

[16] See Miller [1998, 12].

[17] See Miller [1998, 12]. As a matter of fact, developing our hypothesis into full-blown conjecture might itself be seen as an interesting challenge for *FHS*. In this regard, it seems to us that reflecting on the notion of *simulation* might represent a first step in meeting this challenge. We believe that it is an interesting open problem of *FHS* to develop and transform our hypothesis into a proper conjecture. See Paronitti [2008].

[18] Graziani believes that neglecting these principles would encourage building inaccurate and mistaken models and paradigms. Many of these criteria/principles were inspired by Svoboda and Peregrin [2012]; Peregrin and Svoboda [2013]. See also Baumgartner and Lampert [2008]; Baumgartner [2010].

[19] For a more detailed analysis we refer the reader to Graziani [2007; 2014; 2016a; 2016b].

plausibility of what we take to be the fundamental one might nonetheless represent a good start.

It will be necessary, before stating our principle, to clarify the use we make of a couple of terms. First, we accept the following characterization of the expression "scope of the system", given by Svoboda and Peregrin [2012, note 8]:

> We assume that each logical system has been conceived with the goal of accounting for the behavior of a certain part of the logical vocabulary of natural language and the arguments that hold in virtue of this very vocabulary. Classical propositional logic focuses on the behavior of the well known connectives, classical predicate logic adds the basic quantifiers to this and modal logic further adds a certain modal vocabulary, etc. The intended scope of the system is then constituted by the arguments that are correct solely in virtue of the specific kind of vocabulary that the logical system is supposed to capture.

Second, we do not mean to suggest that an historian of science, while including *FHM* (*FHS*) within his bag of tools, should not avail himself of methods stemming from other approaches to his discipline in order to clarify notions that he deems *important*. Accordingly, the notion of 'importance' relevant for our purposes will be the one that best accommodates the results reached by different approaches.[20]

We are now in the position to state our principle:

Principle of Expressivity:
Formal system S is adequate to render the arguments of an informal theory T if and only if the intended scope of S allows a full expression of all the important elements of the grammatical structure of the T sentences and its inferential structure.

In order to clarify the nature of this principle[21], we will now consider its application to the case of the meaning of Euclid's postulates.

2 Questioning the interpretation of Euclid's Postulates

After having provided, in the present section, a detailed analysis of what we take to be the meaning of Euclid's Postulates, we will explain, in the next section, how an adequately chosen and correctly used formal tool might help us in formalizing this meaning.[22] Some of our claims concerning the meaning of Euclid's Postulates will undoubtedly sound surprising to some. A similar reaction, we believe, has two possible explanations. While indeed,

---

[20] An example of this dialectical process will be given in the following paragraphs.

[21] A more general analysis will be presented in Graziani [2016a; 2016b].

[22] As we shall see, the so called *configurational dimension* of the postulates has often been hidden, in the past, by the use of inadequate interpretations and formalizations.

on the one hand, very little efforts have been made to present a coherent picture the most recent literature on the topic, the very question concerning the proper meaning of Euclid's postulates, on the other hand, has not yet been answered once and for all. As a consequence the nature of the Postulates, as it has been observed, has become one of those many "outdated pieces of luggage which we go on carrying around largely because we have forgotten about their existence".[23] Besides advocating the usefulness of formal instruments for historical-mathematical investigations then, a collateral intent of the present paper will be that of providing a critical survey of the most recent literature on the topic and to single out its main themes.

Euclid's *Elements*[24], as it is well known, while not being the first work of its kind, are certainly the first attempt at presenting much of the knowledge about elementary mathematics developed up to his time. Many of the fundamental notions on which it relies are to be found in book I. Amongst these are *Definitions (horoi)*, *Postulates (aitemata)* and *Common Notions (koinai ennoiai)*. After characterizing such notions, Euclid moves on to the statement and corresponding solution/proof of 48 propositions, which are divided into two kinds: those which describe a task (*problems*), and those which make an assertion (*theorems*).[25]

The interpretation of Euclid's Postulates has traditionally been studied in the context of the *existential interpretations of geometric constructions*, i.e. *the idea that Greek mathematicians would use geometric constructions in order to prove the existence of the constructed figures*. This view, which is usually associated to the name of H. G. Zeuthen[26], seemed initially to be both logically and historically reliable. Nonetheless, works due to A. Frajese [1950], E. Stenius [1978], W. Knorr [1983; 1993] and O. Harari [2003] (to mention just the most notable)[27], have showed its limits, thereby con-

---

[23] Netz [2000, 14].
[24] Euclid [1990].
[25] See Euclid [1990].
[26] Zeuthen [1896].
[27] To be more specific, Frajese has provided a more coherent constructive interpretation of the Euclidean postulates; Stenius has put forward a subtle and insightful epistemological reading of the procedures of geometrical constructions; Knorr has showed that not all ancient geometrical constructions were driven by the need of existential proofs and at the same time that Greek mathematicians were able to tackle existential problems in many other ways as well, for instance by means of postulates, tacit assumptions (such as geometrical continuity) and existential theorems (and also constructions); Harari has contributed to undermine the historical credibility and theoretical plausibility of the standard existential interpretation (Zeuthen) of geometrical constructions by insisting on the difference between the Aristotelian notion of 'being' and the modern notion of 'existence'. She also stressed further differences between the Aristotelic and Euclidean notion of 'geometrical object', between the kind of reasoning that Aristotle applies to geometrical issues and the proofs/solutions that Euclid gives in the *Elements*. By doing

tributing to a more adequate understanding of the Euclidean geometrical constructions and hence of the Postulates themselves.[28]

When we analyze the meaning of the five Postulates we can see, at first, that the first three Postulates directly express the *possibility*, or a *required (postulated) ability, to execute specific constructions*, namely segments, extensions of finite straight lines, and circles. For this reason, the Postulates were often taken to refer implicitly to the usage of ruler and compass. It must be kept in mind, though, that no mention of such instruments is to be found in Euclid, and that a strictly instrumental description would require much more than what is actually stated by the Postulates, which, in our opinion, express rather *idealized capabilities*. Euclid presents indeed the Postulates as *minimal assertions* which can be enriched instrumentally by further propositions.[29]

Postulates IV and V, according to Zeuthen's interpretation, would also be constructive in nature: while indeed the former would establish the uniqueness of the extensions of finite straight lines required by Postulate II, the latter would express the condition under which only the intersection point of two straight lines can be built. However, Zeuthen's interpretation of the Postulates does not seem convincing: the reason why Postulate II should have been completed by means of postulate IV, while the other postulates didn't, for instance, is left without explanation.

From an opposite point of view, Attilio Frajese reduces both the meaning of such Postulates and their constructive value to *the problem of the equality of figures*. Here is how he envisions the situation:

> Those geometrical figures which for Plato should be above all object of contemplation are instead for Euclid object of study and of geometrical considerations. But, in order for the figure themselves to become object of study, they first need to be compared to each other, i.e. a link between them needs to be built. In modern mathematics the concept of correspondence produces links, so the elements of a set are connected to each other by means of the structure that the set receives, i.e. by means of operations defined with such properties. For Euclid, figures can be connected by means of constructions and other means that one must be able to use: requirement which, according to us, would be made in the Postulates. So in Postulate I any pair of points would be connected by means of a segment of straight line; the extensions of finite straight lines (Post. II) would also allow us to reach (connect) also the regions which are farther in the plain. A special link between the straight lines is their intersection: the conditions for this intersection to happen

---

so he contributed to clarify the distance between the modern concept of 'existence' and the meaning of the Euclidean geometrical constructions. See bibliography.

[28] A new and interesting analysis can be found in Marco Panza's works mentioned in the bibliography.

[29] For example Proposition I,2 and Proposition I,3 enrich Postulate III by completing it and by allowing the shifting of the segment, which is necessary to give the Postulate an instrumental reading. For an exact analysis see Frajese [1950] and Sidoli [2004]. See Euclid [1990].

are stated in Postulate V. But in those conditions, equality and inequality considerations come into play (comparisons which also connect the figures). Postulates III allows to recognize the equality (and also the inequality) between segments (the circle is a figure which allows for example to recognize that two segments are equal if they are radii of the same circle or if they can be reported to the radii. We shall see that in Proposition 2 and 3 of Book I, Euclid adds something to what is allowed by Postulates III, and explains how to shift the segments. For angles on the other hand one cannot generally recognize the equality by constructively executing the shift as it is possible with segments. Euclid will resort (though somewhat unwillingly and exceptionally) to a kind of mechanical transportation (a real mechanical movement) in Propositions 4 and 8 of Book I. But for at least one kind of angles (the straight ones) Euclid can postulate equality; this is what he does in Postulate IV. Between the straight angles, in other words, wherever they be located in the plane, some kind of remote connection is made: we could call it a radio connection, or wireless, as opposed to the wired connection i.e. with the straight lines of Postulates I and II, and with arcs of circles in Postulate III.[30]

The analyses attempted by many recent studies[31] tend to remain within the boundaries originally drawn by Zeuthen. We accept at the same time Frajese's interpretation as an integration of Zeuthen's. The latter can be seen as a *configurational (relational) interpretation* of the Postulates and of geometrical constructions in general. By "configurational interpretation" we mean an analysis of the solutions of mathematical problems and of the proofs of mathematical theorems which focuses not only on the deductive connection between mathematical propositions (or on the deductive leap from the axioms to the proposition to be proved)[32] but also and especially on the functional connections among the elements of the mathematical propositions contained both in the solutions and in the proofs.[33] The mathematician would therefore be expected to work on the configurations of objects by connecting their elements to each other by means of *interdependencies* i.e. through the *construction of links highlighted by means of geometrical constructions*. From this point of view, Euclidian Postulates would display a very refined constructive dimension: they express *epistemic capacities* relying on which one can construct *geometrical links* between objects which satisfy certain conditions. The act of constructing, in this case, would be

---

[30]Frajese [1950, 302].
[31]See, for instance, Mäenpää [1993; 1997]; Netz (2000); Sidoli [2004]; Harari [2003]; Miller [2008]; Mumma and Avigad and Dean [2009]; Panza [2011; 2012].
[32]This is the "propositional interpretation" or "directional interpretation". Hintikka and Remes [1976] and Mäenpää [1993; 1997] introduced this terminology. Cornford [1932]; Robinson [1936], for instance, present a propositional interpretation within their studies on the ancient method of analysis and synthesis. A Propositional Interpretation of Euclid's *Elements* is also to be found in Hilbert [1899; 2000]; Blumenthal [1961]; Hartshorne [2000].
[33]According to the configurational interpretation, for instance, the ancient method of analysis would be a study of the functional dependencies in a mathematical configuration which encompasses known as well as unknown constituents. See Mäenpää [1993; 1997].

more similar to *producing evidence for the existence* of those geometrical links, than to generating the links themselves.

In a recent study[34], Harari has shown that Euclid considers two different aspects of geometrical elements, namely *material* or *quantitative aspects*, on the one hand, and *positional* or *qualitative aspects*, on the other. In Book I of the *Elements*, for instance, Euclid gives two different characterizations of the same geometrical entities: while in Definition 1.1 a point is defined as "that which has no parts", definition 1.3 tells us that "the extremities of a line are points". Similarly, in the case of lines, we find the two characterizations "breadthless length" and "extremities of a surface". Double characterizations can also be found with respect to other entities. These definitions can therefore be classified either as (1) definitions which determine quantitative aspects of geometrical objects (the former ones), or as (2) definitions which determine qualitative aspect of geometrical objects (the latter ones). The definitions in the former group distinguish measurable from non-measurable aspects and refer to the notion of divisibility in order to determine the measurable aspects of the spatial object: a point would be a non-measurable entity because it has no parts which we can measure/quantify; a straight line would be measurable with respect to its length, but not to its width. The definitions in the latter group, in turn, are aimed at characterizing those aspects that were left out by the definitions of the former, i.e. the non-measurable aspects. In the case of the line, for instance, points are said to be the limits of the quantitative object. The definitions of the two groups correspond therefore to the distinction between *material aspect* (such as divisible or measurable aspects) and *positional aspects* (considered as limits of the measurable figure) of geometrical figures. The priority of material/quantitative aspects over positional/qualitative ones, we believe, shows that Euclid (as opposed to the Pythagorean tradition[35]) does not consider spatial positions as properties of objects, and that the notion of geometrical space, according to him, is not presupposed by geometrical objects. This means, in particular, that the Euclidean space would not be given, but rather *constructed* in the sense of being *made evident*. *It would be a space where geometrical relations and geometrical objects are generated by building links*[36] *amongst, or by limiting, divisible magnitudes (materials)*. Geometrical constructions then, according to our construal, would be best

---

[34] See Harari [2003, 18-19].

[35] It might be useful to recall that a point, according to the Pythagorean definition, would be "a unit having position".

[36] We hold this reading to be consistent with the absence of a specific point-construction postulate. Points, as a matter of fact, do not construct links, they don't have parts and they are limits of other geometrical objects. They have to be assumed and can be obtained indirectly either from the fifth Postulate or from Problems (for instance, I,1).

seen as *epistemic procedures* (i.e. as the acts of building links and of limiting divisible magnitudes) by means of which we can *construct* geometrical objects that satisfy certain conditions. By 'construct' we mean *to supply evidence for the existence* of objects of a specific kind instead of generating them ontologically (positing them as new ontological entities). The Postulates of construction would therefore be interpreted more correctly as stating *basic epistemic capacities*, or *capacities to execute basic procedures*.[37]

What we think emerges from the above investigation is the fundamental *configurational (relational) nature* of the geometrical figures as well as of the mathematical procedures by means of which we prove theorems and solve problems. A geometrical figure, in other words, is not seen from the point of view of its essential attributes (as it was the case in Aristotle), but rather from that of its relationships with other elements of the spatial configurations it belongs to. According to our construal then, proofs and solutions should not be considered *only* in connection with some steps of a deduction from primitive elements to the sought conclusion but rather, and *above all*, in connection to the investigation of specific types of *geometrical configurations*; and the search for proofs and solutions should be conceived, in particular, as a study of the *(functional) interdependencies among the geometrical elements which belong to the relevant configuration*.

The above interpretation offers strong support to the claim, already made by Harari, that geometrical constructions would play a double role in the development of Euclidean propositions: (A) they either serve as a measuring tool by means of which quantitative relations *are deduced*; or (B) they serve as an instrument by means of which it becomes possible *to exhibit* qualitative relations, i.e. the order or the position of geometrical figures. While the deductive employment of the constructions serves the only purpose of *making explicit* a content that is already given in the *ekthesis*, in their second role constructions act as instruments capable of *further developing* the content given in the *ekthesis* by placing the different geometrical entities in different spatial relationships.

We can take, as an example, the Euclidean proof of theorem I.5: "in isosceles triangles the angles at the base equal one another, and, if the equal straight lines are produced further, then the angles under the base equal one another".[38]

In the *apodeixis*[39], Euclid's proof leads from conjunction (1) "line AF is

---

[37] See Stenius [1978]; Mäenpää and von Plato [1990].
[38] See Euclid [1990 vol. I, 204-208]. We are here following Harari [2003, 20-21].
[39] The Euclidean proofs and solutions have forms that can be easily exposed in a schematic way: *Protasis* (Enunciation); *Ekthesis* (Exposition); *Diorismos* (Specification); *Kataskeue* (Constructions); *Apodeixis* (Demonstration); *Sumperasma* (Conclusion). On this partition see the accurate analysis in Acerbi [2011].

equal to line AG and line AB is equal to AC" to conjunction (2) "the angle ACF is equal to angle ABG and angle AFC is equal to angle AGB". It is important to notice that, while both propositions rest on the same construction steps which precede the proof, the role played by these construction steps in the tacit inferences leading up to the two propositions is different. Proposition (1), as a matter of fact, is nothing but a reiteration of the content, which is already at hand in the initial setting-out and in the construction step. By establishing the relation of equality between line AF and AG, this construction step serves measurement purposes, in that it determines the quantity of line AG in comparison to the quantity of line AF. Proposition (2), by contrast, does not rest solely on the content, which is given in the initial setting-out, in the construction steps and in the first congruence theorem. Rather, in order to derive proposition (2) we are forced to give a new meaning to the ingredients of the figure. That is to say that the application of the first congruence theorem requires an act of visualization in virtue of which a configuration of lines is treated *as* a figure of a certain sort. For instance, prior to the proof, line AB is considered to be a side of a triangle, while line BF is regarded as a part of the additional line BD. The application of the first congruence theorem to this proof requires an act of visualization whereby line AB and line BF are treated as sides of a triangle. This act of visualization introduces the triangle AFC, thereby conferring a new meaning to the other components of the configuration: line AC does not play the role of the edge of triangle ABC anymore, and line FC turns into the base of a triangle. The point is that, in the course of *apodeixis*, the meaning of most components of the spatial configuration undergoes a modification. Line BF, for instance, is 'detached' from the whole AF and it is treated as a side of the triangle FBC. The introduction of triangle FBC, in turn, modifies the meaning of the other components of the configuration. Indeed line FC, which served as the base of triangle FAC in the first part of *apodeixis*, turns now into one of the edges of triangle BFC. As a consequence line BC, which originally served as the base of the triangle ABC, has turned into the base of triangle BFC by the end of *apodeixis*.

3 Questioning the formalization of Euclid's Postulates

A widespread interpretation of the Postulates[40] construes them as *existential assertions in an ontological sense* and formalizes them by means of *existentially quantified propositions or existential axioms*[41]. Now, if we

---

[40]See Hilbert [1999]; Mueller [1981]; Hartshorne [2000]. About Hilbert see von Plato [1997].

[41]Proclus [1873, 77-82; 201, 5-9] introduced the idea of an existential interpretation of geometrical problems by using Aristotelian terminology. In modern and contemporary

consider the construction postulates as a proposition of the kind $(\exists x)P(x)$, which states that at least one of the objects of a given domain has the property P, then we should read, say, the first Postulate as stating that between any two points there is a straight line (within the domain of straight lines) which has the property of passing through those two points. However, as we have seen, this does not seem to reflect the assertion made by the first Postulate, which in fact states something about *what we can do* in the sense of *epistemic capacities to execute procedures*.

So far our considerations seem to support the stance advocated by Mäenpää and von Plato[42] who, by *strengthening the relational interpretation of the Postulates*, achieve a *functional interpretation* of them. According to the same view, for instance, the first Postulate would *highlight, by exploiting an ability, the existence of a functional relationship*, namely a function which takes two points as input and gives their connection through a finite straight line as output. A similar reading, moreover, suggests that the first Postulate is much more complex indeed than a mere axiom of existence, insofar as it does not explicitly talk about the existence of the set of all segments nor it implies any ontological attribution of existence but, to put it in modern terms, it reminds us rather of the need of some kind of *Introduction Rule* for such a set by showing us the way in which its typical elements must be found. This means that, in the case of some entities, no statement of existence in the ontological sense is given, as by so doing one would assume too much. The existence of a link between ontological existence and our ability to investigate a given domain is immediately assumed by the act of showing (presenting) the way in which its elements can be constructed or produced and by establishing when two elements are equal.[43]

In order to delve deeper into the general idea we just sketched and to evaluate the application of the first part of the Principle of Expressivity, we might begin by noticing that the interpretation of construction postulates, solutions to problems and proofs of theorems proposed above is still in need of something like a logic or a logical system capable of adequately representing their configurational dimensions. Now, in order to devise a logical system capable of representing geometric constructions[44], we need to rely on an approach similar to Kolmogorov's problem interpretation of logic[45]. Mäenpää

---

times, mathematicians generally consider the interpretation of problems as existential propositions and theorems as universal propositions.

[42]See Mäenpää and von Plato [1990].

[43]See note 47.

[44]See Mäenpää and von Plato [1990] for an accurate analysis.

[45]Kolmogorov [1932]. Logical propositions, according to Kolmogorov, can be understood either as problems or as tasks. Here is an example: Suppose that A, B, C, ... are single elementary problems which can be combined into a more complex one. Now, the

and von Plato [1990], for instance, use the language of the *Constructive Type Theory*[46] (*CTT*) because Predicate Logic, which would be normally used, is not sufficient for a natural logical description of the configurational interpretation, nor of the construction postulates, nor of the auxiliary constructions. They use Type Theory because it enriches Predicate Logic with a functional hierarchy which proves capable of capturing exactly, on the formal level, the relevant Greek informal notion, thereby allowing a unified description of both propositional and configurational dimension. As a matter of fact, in Martin-Löf's Theory we "can interpret a proposition as the problem of finding a proof of the proposition, and a problem as the proposition that there is a solution to the problem"[47] and we also have a form of judgement (assertion) which makes the solutions to problems explicit: $a : A$ reads *a is a solution to problem A*. Indeed, according to the *propositions as type principle*[48], problems, propositions and sets are conceptually identified in Type Theory (they are equal *types*). As a consequence, the elements of sets, proof of propositions and solutions to a problems are treated as identical in Martin-Löf's Type Theory, to the effect that we can express, for instance, the judgment form $a : A$ in the following way: *if an object a of type A exists, an A exists in the primitive sense*. *CTT*, moreover, makes use of different kinds of interrelated proof theoretical rules, within a type theoretical setting, which, according to our Principle of Expressivity, can formalize the configurational nature of the *Elements'* geometry:

> Each set of geometric objects is given *formation rules* which show how a set can be formed; from other sets or as simple, *introduction rules* which give the meaning of the set by showing how canonical objects of the set are formed and when two such objects are equal, elimination rules which show how to define functions over the set defined in the introduction rules, and *equality rules* which show how the functions defined by the *elimination rules* operate on the canonical objects of the set defined by the introduction rules, that is, how they are evaluated. The geometrical rules are used together with the other rules of Intuitionistic Type Theory.[49]

---

complex problem $(A \wedge B)$, for instance, will be solved by solving both A and B; $(A \vee B)$ by solving at least one of A and B; $(A \rightarrow B)$ by reducing the solution of B to that of A; $(\neg A)$ by reducing an impossible task to A

[46]See Martin-Löf [1984]; Sommaruga [2000]; Granström [2011].

[47]Mäenpää and von Plato [1990, 277].

[48]Formally: $Set = Prop : Type$. The judgement form $a = b : A$ means that a and b are equal objects of type A. For example: if A is a *Set*, $a = b : A$ means that a and b are equal elements of A. A *Set* is defined by giving exhaustive conditions for forming its elements and its equal elements. A *Type* is defined by explaining what it is to be an object of the type and when two objects are equal. See Martin-Löf [1984] for further details.

[49]Mäenpää and von Plato [1990, 281]. The fourth Postulate can be viewed as dealing with the equality of two canonical elements of *Straight Angle:Set*. The constructions of such equality are given in problems and theorems: Euclid, as we said, presents the Postu-

If then, as it seems correct, the existential connotation of the Postulates, and of geometrical construction in general, is indeed more primitive than $(\exists x)(Px)$ and, as Mäenpää points out, it expresses something similar to the judgment form $a : A$ in CTT, (*if an object a of type A exists, an A exists in the primitive sense*); and if, as it also seems to be the case, we need something like Kolmogorov's problem interpretation of logic in order to formalize the logic of the *Elements'* construction procedures (and its configurational point of view), then, following Mäenpää and von Plato, we can use CTT as a tool of structural formal analysis of our subject.

For instance, by means of an *Introduction Rule*[50] for segments, we might formalize the first construction postulate as follows

$$\frac{a{:}Point \quad b{:}Point}{l(a,b){:}Segment} \; [\text{lIntroduction}]$$

where a and b are points and their connection by means of a segment is expressed by $l(a, b)$. This inference introduces the basic geometric function l which has geometric constructions as arguments and value. By considering the Principle of Expressivity, we can now observe that such rule expresses the first Euclidean Postulate more clearly than an *existential instantiation rule* could ever do.

As a matter of fact, the usual natural deduction rule:

$$\frac{\exists x \; B(x)}{B(a)}$$

reduces it to a mere exemplification (i.e. a way of representing a universal concept by introducing a specific corresponding instance)[51] and the following rule:

$$\frac{\vdash Point(a) \quad \vdash Point(b)}{\vdash Segment(l(a,b))}$$

where some individual properties are inferred from others, does not conceal (reveal) any compositional and constructive dimension by eclipsing any configurational dimension.

By contrast, the Mäenpää-von Plato *lIntroduction Rule establishes both a*

---

lates as minimal assertions which can be enriched instrumentally by further propositions. See Graziani [2007; 2014].

[50]Mäenpää and von Plato [1990]; Mäenpää [1997]. The Formation Rule is: *Segment:Set*.

[51]Hintikka and Remes propose this interpretation in Hintikka and Remes [1976].

deductive connection between premises and conclusion, and functional dependencies between the constructions in the premises and the construction in the conclusion.

This means in particular that, in developing problems and theorems, we can have both deductive rigor and the ability to express deductions within a configuration made of functional links among data, conditions, auxiliary constructions, and the thing sought.

The same line of reasoning leads us to the conclusion that the widely shared interpretation of problems as existential propositions (and, correspondingly, of theorems as universal propositions) is conceptually inappropriate. A problem, in the Greek sense, aims at constructing a certain sought object by putting it in specified relations with other given objects. Its linguistic formulation reflects this by relying on a typical infinitive form which was normally used in ancient Greek in order to express the need to fulfill a requirement. A theorem then, according to the same construal, will be best thought of as the requirement to demonstrate a property of a given object. It seems therefore plausible to maintain that a problem was standardly thought by a Greek mathematician as being composed of three parts, namely the *given*, the *thing sought* and the *condition*.[52] This means that, in order to solve a problem, we need to construct the thing sought from the datum and we subsequently have to prove that the condition holds. The intrinsic interest of the thing sought is what distinguishes a problem from a theorem, which in turn does not have a thing sought, and in order to prove which, a mathematician needs to prove the condition for the datum.

It is also important to note that the Greeks did not distinguish the thing sought from the conditions. Indeed, they used the term *zetoumenon* to refer to a combination of both.[53] This has the drawback of creating ambiguity. In this regard, the use of correct formal tools (CTT), whose choice is made by taking the Principle of Expressivity into account, has the further merit of highlighting this distinction.

As a matter of fact, the fundamental difference in content between the stating of a theorem and that of a problem, according to Petri Mäenpää, can be formalized in CTT as

$$(\exists y : B(x))C(x,y) : Problem \ (a : A)$$

where $a : A$ is the *datum*; $y : B(x)$ is the *thing sought*; and $C(x,y)$ is the *condition*.[54] This means that, in order to solve a problem we have to con-

---

[52]A problem need not necessarily display all three of them. On this partition see the accurate analysis in Mäenpää [1993]; Acerbi [2011].

[53]Mäenpää [1993; 1997] was the first to note this.

[54]As special cases we have $(\exists y : B)C(y) : Prop$ (when the datum is missing); $C(x) : Prop \ (x : A)$ (when the thing sought is missing); $B(x) : Set \ (x : A)$ (when

struct a $y : B(x)$ from $x : A$ and we have to prove that $C(x, y)$ holds.[55] It is important to note that the existential proposition is used here to express a problem with a condition. Problems, in other words, are not identified with existential propositions.

Similarly, a theorem can be formalized as follows $C(x)$ $(x : A)$, which is a borderline case of a problem, since it lacks the thing sought. In order to prove a theorem then, a mathematician will try to prove $C(x)$ for $x : A$.[56] It must be observed that a theorem can also be an existential proposition, for instance in the case in which the condition is itself an existential proposition. We can find examples of existential theorems in Euclid's *Optic* (Theorems 37 and 38).

As we have tried to show, both CTT and the Mäenpää-von Plato formalization allow us to express many ideas which undeniably belong to the cultural heritage of Ancient Mathematics and which other formalizations cannot express (or worst run the risk of hiding). By considering previous analyses of the nature of Euclidean constructions[57] it is possible to provide[58] a set of axioms and rules [named *EPH*] which *follow our principle and which, by modern standards, prove themselves sufficient to provide a foundation for Euclid's Plane Geometry*. *EPH*[59] have the merit of making explicit many concepts that Euclid implicitly included in his system and relegated to an intuitive understanding. The basic structure of this axiomatization includes a number of basic constructive concepts, the general properties that these basic concepts have, the realization of some ideal situations through construction postulates, the properties and uniqueness of constructed objects, and the compatibility among the various concepts and constructions.[60] The formal theory underpinning this axiomatization is CTT.

---

the condition is missing).

[55] An example of problem is the first Proposition of Euclid's *Elements*: to construct an equilateral triangle on a given line segment. Here the given object is a line segment, the sought object is a triangle, and the conditions are that the triangle must be both equilateral and construed on the line segment.

[56] An example of theorem is Proposition I, 32 of Euclid's *Elements*, which states that the angle sum of a triangle equals two right angles. Here the given object is a triangle and the condition is that the sum of its angles be equal to two right angles.

[57] See also Pambuccian [2008].

[58] See Graziani [2007; 2014].

[59] Today there are several studies on this topic, all developed independently from each other. See, in particular: Hartshorne [2000]; Miller [2008]; Mumma [2006]; Mumma and Avigad and Dean [2009]. An attempt at comparing all these different approaches will be presented in Graziani [2016a: 2016b].

[60] von Plato (see bibliography).

4 Conclusions

It is not possible here to either present *EPH* or to show the role played within it by the fundamental principles listed in the introduction.[61] Our intent was just that of showing small portions of the formalization, which we held sufficient to grasp, although in a general and intuitive way, the role of the Principle of Expressivity. We have tried to show the connections amongst some pieces of recent literature on the meaning of construction procedures in Euclid's *Elements* and we have argued in favour of their configurational interpretation. Subsequently, following the Principle of Expressivity, we have argued in favour of a type-theoretic system which might help formalizing them. Other formal tools, as we have seen, fall short of making explicit many of the concepts Euclid relied on in his system. Although they have appeared natural to many scholars, some interpretations and formalizations, as we have shown, are indeed inadequate interpretations and formalizations of Euclid's construction procedures, proofs and solutions.[62] In the light of the above considerations we are now in the position to maintain that the answer to the question "whether formal logic can aid our analyses of ancient mathematics" will be affirmative, if the use of formal logic is informed by principles similar to the one we have proposed, and negative otherwise.

To conclude, we believe that this very brief analysis illustrates how *properly chosen and used*[63] formal tools can help the study of ancient mathematics. We regard the exploration of *FHM* (*FHS*) principles not only as a new and promising domain of application of logic, but also a philosophically relevant subject matter.

BIBLIOGRAPHY

ACERBI F., 2011, Perché una dimostrazione matematica greca è generale, in F. Repellini, E. Nenci (eds.), *Atti del Workshop "La scienza antica e la sua tradizione"*, pp. 25-80, Milan, LED.

BAUMGARTNER M. and LAMPERT T., 2008, Adequate formalization, *Synthese*, 164, pp. 93-115.

BAUMGARTNER M., 2010, Informal Reasoning and Logical Formalization, in *Ding und Begriff*, S. Conrad and S. Imhof (eds.), Frankfurt, Ontos Verlag.

---

[61] See Graziani [2007; 2014; 2016a; 2016b].

[62] See notes 11, 31, 39.

[63] As we have seen in our case study, we have to consider both the expressive capabilities of the language and the dynamics of the deductive apparatus chosen for the formalization. Obviously, these choices should depend on the content of ancient mathematical texts and their logical syntax (see Acerbi [2011]). The use of formal tools and analysis of ancient mathematical texts should therefore be in a relationship of constant mutual growth.

BEESON, M., 2009, Constructive geometry, in: Arai, T. (ed.) *Proceedings of the Tenth Asian Logic Colloquium, Kobe, Japan*, pp. 1984. World Scientific, Singapore.
BEESON M., Ts, Foundations of Euclidean Constructive Geometry, http://www.michaelbeeson.com/research/papers/Constructive GeometryLong.pdf
BEESON M., 2015, A constructive version of Tarski's geometry, *Annals of Pure and Applied Logic*, 166(11):1199-1273.
BERTATO F. M., 2014, Sobre as formalizações sillogísticas dos Elementos, effettuadas por Herlinus, Dasypodius, Clavius e Hérrigone, in S. Nobre, F. M. Bertato, L. Saraiva (eds.) *Anais/Actas do 6 Encontro Luso-Brasileiro de História da Matemática*, Sociedade Brasileira de História da Matemática, Natal.
BLUMENTHAL L. M., 1961, *A modern view of geometry*, New York, Dover Publications.
CALOSI C. and FANO V., 2015, Divisibility and Extension, *Acta Analytica*, 30(2): pp. 117-132.
CORCORAN J. (ed. by), 1974, *Ancient Logic and its Modern Interpretations*. D. Reidel Publishing Company, Dordrecht-Holland.
EUCLID, 1990, *Euclide, Les Éléments*, translated into French, with introduction and commentary by B. Vitrac, 4 vols., Presses Universitaires de France, Paris.
FEFERMAN S., 2012, And so on ...: reasoning with infinite diagrams, *Synthese*, 186, 1, pp. 371-386.
FRAJESE A., 1950, Sul significato dei postulati euclidei, *Scientia*, vol.LXXXV, N.CDLXIV-12, pp. 299-305.
GRANSTRÖM J. G., 2011, *Treatise on Intuitionistic Type Theory*, Berlin, Springer.
GRAZIANI P., 2007, *Analisi Strutturale e Fondazionale della Geometria del Piano di Euclide*, PhD Thesis, University of Rome "La Sapienza".
GRAZIANI P., 2014, A structural and Foundational Analysis of Euclids plane geometry: the case study of continuity. In V. Fano, F. Orilia and G. Macchia (ed. by), *Space and Time. A Priori and A Posteriori Studies*. DeGruyter-Ontos Verlag, pp.63-106. Berlin/Boston.
GRAZIANI P., 2016a, Contemporary foundations of Euclid's geometry, in preparation for *Isonomia*.
GRAZIANI P., 2016b, Criteria for logical formalization in Formal History of Sciences, in preparation for *Isonomia*.
HARARI O., 2003, The concept of Existence and the Role of Constructions in Euclid's Elements, *Archive for History of Exact Sciences* N. 57, pp. 1-23.

HARTSHORNE R., 2000, *Geometry: Euclid and beyond*, New York, Spinger-Verlag.
HILBERT D., 1899, *Grundlagen der Geometrie*, (eleventh edition, 1972, B. G. Teubner, Stuttgart), Teubner, Leipzig.
HILBERT D., 2000, *David Hilbert's Lectures on the Foundations of Geometry 1891-1902*, M. Hallett and U. Majer (eds.), Berlin Heidelberg New York, Springer.
HINTIKKA J. and REMES U., 1974, *The method of analysis: Its Geometrical Origin and Its General Significance*, Reidel, Dordrecht.
HINTIKKA J. and REMES U., 1976, Ancient geometrical analysis and modern logic, in R. S. Cohen et al. (eds.), *Essay in memory of Imre Lakatos*, Reidel, Dordrecht, pp. 253-276.
KNORR W. R., 1983, Construction as existence proof in ancient geometry, *Ancient Philosophy*, N.3, pp. 125-148.
KNORR W. R., 1993, *The ancient tradition of geometrical problems*, New York, Dover Publications.
KOLMOGOROV A., 1932, Zur Deutung der intuitionistischen Logik, *Mathematische Zeitschrift*, 35, pp. 58-65.
LUENGO I., 1995, *Diagrams in Geometry*, Ph.D. thesis, Indiana University.
LUKASIEWICZ J., 1957, *Aristotle's Sillogistic from the standpoint of modern formal logic*, Oxford, Oxford University Press.
MÄENPÄÄ P. and VON PLATO J., 1990, The logic of Euclidean construction procedures, pp. 275-293 in L. Haaparanta, M. Kusch e I. Niiniluoto (eds.), *Language, Knowledge, and Intentionality. Perspectives on the Philosophy of Jaakko Hintikka*, Acta Philosophica Fennica 49 (The Philosophical Society of Finland, Helsinki).
MÄENPÄÄ P., 1993, *The art of analysis. Logic and history of problem solving*, PhD Thesis, University of Helsinki.
MÄENPÄÄ P., 1997, From backward reduction to configurational analysis in M. Otte and M. Panza (eds.) *Analysis and Synthesis in Mathematics*, Boston Studies in the Philosophy os Science 196, Reidel, Dordrecht, pp.201-226.
MÄENPÄÄ P., 1998, A reductive sequent calculus for proof search in type theory, available at http://www.helsinki.fi/filosofi/personal/jvp/maenpaa.pdf
MARTIN-LÖF P., 1984, *Intuitionistic Type Theory*, Bibliopolis, Naples.
MILLER N., 2001, *A Diagrammatic Formal System for Euclidean Geometry*, Ph.D thesis, Cornell University.
MILLER N., 2006, Computational complexity of diagram satisfaction in Euclidean Geometry, *Journal of Complexity*, 22(2), pp. 250-274.
MILLER N., 2008, *Euclid and his Twentieth Century Rivals: Diagrams in*

the Logic of Euclidean Geometry, CSLI, Stanford.
MUELLER I., 1981, *Philosophy of mathematics and deductive structure in Euclid's Elements*, MIT Press, Cambridge.
MUGNAI M., 2010, Logic and Mathematics in the Seventeenth Century, in *History and Philosophy of Logic*, vol. 31, 4, pp. 297-314.
MUMMA J., 2006, *Intuition Formalized: Ancient and Modern Methods of Proof in Elementary Geometry*, PhD thesis, Carnegie Mellon University.
MUMMA J. and AVIGAD J. and DEAN E., 2009, A formal system for Euclid's Elements, *The Review of Symbolic Logic*, 2, pp. 700-768.
MUMMA J., 2010, Proofs, Pictures and Euclid, *Synthese*, 175, 2, pp. 255-287.
MUMMA J., 2012, Constructive geometrical reasoning and diagrams, *Synthese*, 186,1, pp. 103-119.
NETZ R. (2000), Why did Greek Mathematicians Publish Their Analyses?, in P. Suppes, J. M. Moravcsik, H. Mendell (eds.), *Ancient and Medieval Traditions in the Exact Sciences*, CSLI Publications, pp.139-157.
PANBUCCIAN V., 2008, Axiomatizing Geometric Constructions, *Journal of Applied Logic* 6, pp. 24-46.
PANZA M., 2011, Rethinking Geometrical Exactness, *Historia Mathematica*, 38, pp. 42-95.
PANZA M., 2012, The twofold role of diagrams in Euclid's plane geometry, *Synthese*, 186, 1, pp. 55-102.
PARONITTI G., 2008, *Epistemologia della Simulazione*, LuLu.com.
PEREGRIN J., 2006, Meaning as an Inferential Role, *Erkenntnis*, 64, 1-35.
PEREGRIN J., 2008, What is the Logic of Inference?, *Studia Logica*, 88, pp. 263-294.
PEREGRIN J. and SVOBODA V., 2013, Criteria for logical formalization, *Synthese*, 190: pp. 2897-2924.
PROCLUS, 1873, *Procli Diadochi in Primum Euclidis Elementorum Librum Commentarii*, G. Friedlein (ed.), B. G., Teubner, Leipzig.
RIDOLFI L., 1999, *Una analisi formale della sillogistica modale aristotelica*, Urbino, Graduation Thesis University of Urbino.
SIDOLI N., 2004, On the use of term diastema in ancient Greek constructions, *Historia Mathematica*, 31, pp. 2-10.
SIDOLI N. and SAITO K., 2012, Comparative Analysis in Greek Geometry, *Historia Mathematica*, 39, pp. 1-33.
SOMMARUGA G., 2000, *History and Philosophy of Constructive Type Theory*, Dordrecht, Kluwer Academic Publishers.
STENIUS E., 1978, Foundations of mathematics: ancient Greek and modern, *Dialectica* 32, pp. 255-290.
SVOBODA V. and PEREGRIN J., 2012, Logical Form and Reflective

Equilibrium, M. Peli and V. Punoch (eds.): *The Logica Yearbook 2011*, pp. 191-210,

THOM P., 1996, *The Logic of Essentialism. An Interpretation of Aristotles Modal Syllogistic.* Kluwer Academic Publisher, Dordrecht, Nederland.

VON PLATO J., 1995, Organization and development of a constructive axiomatization, S. Berardi e M. Coppo (eds.) *Types for proofs and programs*, pp. 288-295, Springer-Verlag.

VON PLATO J., 1997, Formalization of Hilbert's geometry of incidence and parallelism, *Synthese*, 110, pp.127-141.

VON PLATO J., 1998, Constructive theory of ordered affine geometry, *Indagationes Mathematicae* 9(4), pp. 549-562.

ZEUTHEN H.G., 1896, Die geometrische Construction als Existenzbeweis in der antiken Geometrie, *Mathematische Annalen* 47, pp. 222-228.

PIERLUIGI GRAZIANI
University of Chieti-Pescara,
Department of Philosophical, Pedagogical and Economic-Quantitative Sciences

MAURIZIO COLUCCI
Yellow Blue Soft, CTO

Coherence and Contradiction
JOHN T. KEARNS

1. Fundamental illocutionary acts

Some supporters of paraconsistent logic — Graham Priest is an example — have attributed the widespread acceptance of the logical principle of (non-) contradiction to Aristotle. They consider this to be one more of Aristotle's many bad ideas, perhaps one of the most influential and long-lasting of his bad ideas. In this paper, I will argue against the view that the principle of contradiction is a bad idea. Properly understood, the principle is obviously, evidently, correct, and every rational being must realize this.

I will approach this topic within the framework provided by *illocutionary logic*, which is the logic of speech acts. When a person speaks, writes, or thinks with words, that person is performing *speech acts*, or *language acts*. Since these acts do not all involve speaking aloud, the expression 'language act' may be more appropriate than 'speech act', but I use these expressions interchangeably. A *sentential* act is a meaningful act performed by using a complete sentence or a sentential clause, and a *statement* is a sentential act which is true or false — at least, it is one that can appropriately be evaluated in terms of truth and falsity.

Some acts performed with sentences constitute *illocutionary acts*. An illocutionary act is the act a person is intending to perform when she speaks, writes, or thinks a sentence, an act like an assertion, a promise, or an apology. When used in performing such an act, a sentential act is said to have a certain *illocutionary force*. The illocutionary act is constituted by performing the sentential act with illocutionary force. Statements themselves can be performed with different illocutionary forces; a statement can be asserted or it can be denied, for example. An *assertive* act is an illocutionary act performed by making a statement, where the illocutionary force concerns the issue of whether or not that statement "fits" the world.

An *assertion* is here understood to be an act of producing, or *performing*, a statement and accepting that statement as being or representing what is the case, or else an act of producing a statement, and reaffirming one's continued acceptance of that statement. Assertion is a fundamental, and inescapable, type of language act. There can be no human language whose speakers don't make assertions, although they might make specialized forms of assertion.

A simple statement predicates an expression of one or more objects. The predicate expression is associated with a criterion, and is truly predicated of objects which satisfy the criterion. Since predicates are intended to "fit" objects which satisfy their associated criteria, simple statements are *designed* for providing true representations of the world. Even complex statements are designed to truly represent the world, although their representing is more complicated. Since statements are designed for representing things as they are, what "comes naturally" to a statement is an act of asserting, or accepting, that statement.

We can say that a statement is internally, or *intrinsically, successful* if it fits the world, though speakers don't always intend for their statements to fit the world. A language user can *reject* a statement rather than accepting it. Strictly speaking, it isn't the statement that is rejected, it is the assertion, or the act of accepting a statement, that is rejected. A denial of a statement *rules out*, or *blocks*, the assertion of a statement, on account of the statement's failing to fit the world. In familiar natural languages, it is common to make a denial by interrupting, or interfering with, the formation of a statement. If we use the sentence:

Nancy is not in Alsace.

to deny that Nancy is in Alsace, the internal 'not' will not function as statement negation in systems of propositional logic. Instead, the 'not' makes explicit the force of denial by "dividing" the act of referring to Nancy from the act of predicating 'is in Alsace' of Nancy.

It isn't possible to both do something and refrain from doing it on the same occasion, in the same respect. It isn't possible, for example, to both open the door to the room and refrain from opening the door. A person could first open the door, and then close it. Or she could refrain from opening the door for a while, and then open it. But she can't do both at once, and nobody thinks that she can. Since it isn't possible to do both things at once, it would make no sense to ask someone to do both things at once. A genuine request is one where the speaker wants the addressee to do what he is asked, or, at least, to try to do it. If a speaker asks someone to both open the door and refrain from opening it, she can't want the addressee to do what he is asked. Such a "request" would not be a genuine request.

It is equally senseless to both accept a statement and rule out its acceptance. A person can do one of these, then change her mind and do the other. But when she changes her mind, she "takes back" the illocutionary act she first performed. She can't really do both. If we say that it is *incoherent* to accept and deny a single statement, and that it is also incoherent to perform assertive illocutionary acts that commit one to accept and deny a single statement, then language users are *rationally required* both to avoid performing incoherent acts, and to eliminate incoherence once they recognize their own acts to be incoherent.

2. Logical theories

Illocutionary logic is the logic of speech acts, or language acts. It is primarily the logic of assertive illocutionary acts, although we can also investigate logical features of those acts which John Searle calls *directives* and *commissives* in [9]. Commands and requests are examples of directives, while promises are the paradigm of commissives. Illocutionary logical theories, at least those theories which I investigate, have two levels, an *ontic* level and an *epistemic* level. A first-level, or ontic-level, theory is a standard logical theory. It has a *framework*, or *frame*, containing an artificial formal language, a semantic account for that language, and a deductive system; in addition to the frame, the theory contains derivations in the deductive system and results established by these derivations. The first-level languages contains what I call *plain* sentences, these are formed with connectives and (sometimes) with quantifiers and other operators. Plain sentences represent *statements*.

In developing a first-level logical theory, it is customary to investigate both truth conditions of statements and semantic features defined in terms of, or determined by, truth conditions — features like implication, logical consequence, and inconsistency. In such a theory, deductive procedures are devised for tracing truth-conditional connections linking statements and sets of statements. Although it is common to speak of these procedures as producing arguments, or proofs, these are not the kind of arguments or proofs that are found outside of logic courses and logic books. I prefer to say that first-level systems give us *deductive derivations* of some statements from others. Genuine arguments and proofs are *speech-act arguments* and *proofs*. These are investigated in second-level illocutionary theories. Premisses and conclusions of speech-act arguments are illocutionary acts rather than statements.

A second-level, or epistemic-level, illocutionary theory is obtained by adding illocutionary force-indicating expressions to the first-level language, and exploring the *rational commitment* which links illocutionary acts. The basic illocutionary force-indicating expressions, or *illocutionary operators*, are the following:

⊢ the sign of assertion
⊣ the sign of denial
⊢ the sign of positive supposition
¬ the sign of negative supposition

The signs for assertion and denial are borrowed from Frege. The sign of positive supposition is my own invention, while the sign of negative supposition is often used for negation, especially in systems of Intuitionist logic — but I am using this sign to indicate an illocutionary force.

If A is a plain sentence of the artificial language, then the *completed* sentences constructed with A are the following:

⊢A, ⊣A, ⊢A, ¬A

Completed sentences represent *assertive illocutionary acts*. These sentences and the illocutionary acts which they represent are the basic elements of second-level theories of illocutionary logic. I have already explained those illocutionary acts which are assertions or denials. A *positive supposition* is an act of temporarily accepting a statement, to explore the consequences of combining this with assertions and denials that a person has made and, possibly, with other suppositions as well. A *negative supposition* is the negative counterpart to a positive supposition, and blocks or impedes the temporary acceptance of a statement.

A second-level deductive system constructs, or represents, speech-act arguments. The premisses and conclusions of speech-act arguments are illocutionary acts. In the second-level deductive system, steps in arguments are completed sentences rather than plain sentences. A speech-act argument might begin with premisses which are assertions or denials, and conclude with an assertion or denial. Suppositions are also employed in speech-act arguments, and are sometimes *discharged* in the course of an argument.

A statement is not, in general, tied to a single person. Different people can make what is essentially the same statement. But an illocutionary act is always some particular person's act. Performing one or more assertive illocutionary acts will *commit, rationally commit*, a person to perform further illocutionary acts. However, Jones' assertion of statement A commits Jones to perform further acts, but has no such consequences for Smith, while Smith's acts commit her, but not Jones. A second-level illocutionary logical theory is a *first-person theory*, it is a theory that some person uses to represent her own acts and arguments. But the logician's job is to devise a *generic* theory that each person can adapt to her own situation. Each person employs the same rules or principles in constructing arguments, for example, but each person's arguments begin with her own assertions, denials, and suppositions, and conclude the same way. A person cannot, for example, properly assert A as a premiss unless she believes A. A person's first-person illocutionary theory contains her own arguments, which may serve to enlarge her knowledge and belief. Because generic logical theories serve a normative function, I construct generic (second level) illocutionary theories for an ideal, or idealized, language user whom I call the *designated subject*, for whom I use feminine pronouns.

Performing some *assertive* acts (assertions, denials, and suppositions belong to the category of assertive acts) will commit a person to perform others. But rational commitment is not simply a relation connecting assertive illocutionary acts. A person can be rationally committed to perform any kind of intentional act, or to remain in a certain state, like that of accepting a given statement. Making a decision to carry out a given action will rationally commit a person to carry out that action. However, rational commitment is not some kind of causal necessity. Rational commitment, when recognized, *motivates* a person to act, but it may not carry the day.

Commitment is either *conditional* or not. A commitment to close the upstairs windows in my house if it rains while I am at home is conditional on its raining while I am at home. But if, on a given Thursday, I decide to buy gas for my car on my drive home from work, this generates an unconditional commitment (but probably not a moral commitment). Coming to accept, or continuing to accept, some statements, and rejecting certain statements, will commit a person to accepting other statements, and to rejecting other statements. Positively or negatively supposing statements will commit a person to supposing others (either positively or negatively). If the person who accepts certain statements, and rejects others, is committed to, say, accept statement A, this commitment is conditional. She is committed to accept A if she has some interest in the matter, and gives it some thought. Although a given person will accept many statements and reject many others, she will be uninterested in exploring the consequences of most of these beliefs and disbeliefs.

For example, asserting that today is Thursday will commit a person to accept that either today is Thursday or it is now snowing in Beijing, but most people would have no interest in that consequence, and would not be inclined either to consider or accept it. Indeed, someone might consider the disjunctive consequence to be irrelevant to the claim that today is Thursday. However, it would be irrational to accept that today is Thursday and refuse to grant that either today is Thursday or it is now snowing in Beijing.

Commitment linking illocutionary acts has some resemblance to implication, but is unlike implication in that commitment is either *immediate* or *mediate*. Immediate commitment is evident to the person who is committed, and who gives the matter some thought. Mediate commitment is constituted by a sequence, or chain, of immediate commitments. If doing X_1 immediately commits a person to do X_2, and doing X_2 immediately commits her to doing $X_3,\ldots$, and doing X_{n-1} immediately commits her to doing X_n, then doing X_1 *mediately* commits her to doing X_n. It is immediate commitment which motivates a person to act, for a person's mediate commitments need not be evident to her.

A genuine argument, or a "real-life" argument, is a speech-act argument. It begins with premises which are assertive illocutionary acts, and concludes with an assertive act. A *deductively correct* speech-act argument traces a chain of immediate commitments linking the premiss acts to the conclusion act. These commitments linking assertive illocutionary acts are not arbitrary, or conventional. Given the meanings and truth conditions of statements that are asserted, denied, or supposed, together with the character of the illocutionary acts performed, the language user's commitments are determined "in the nature of things". Deductive systems in second-level illocutionary theories provide the resources for constructing speech-act arguments from initial assertions, denials, and suppositions to a conclusion which is also an illocutionary act.

If performing some illocutionary acts commits a person to perform another, then the first illocutionary acts *logically require* the further act. It is *incoherent* to assert and deny a single statement, or to positively and negatively suppose a single statement. Illocutionary acts which logically require incoherent acts are themselves incoherent.

3. Negation and denial

In *Begriffsschrift*, Frege says in [1] that in his formal language, the assertion sign can be regarded as a predicate of all judgments (which are, for us, assertions). But this is the wrong way to think of the assertion sign. For a predicate is used to characterize those objects it is predicated of. In asserting a statement, we are *not* characterizing the statement, we are *accepting* it. We *can* characterize the statement, as when we say that it is true, or false, or interesting. But assertion comes first, it is *prior* to acts characterizing the statement as true.

The illocutionary act of denial has a negative character, but this is not an act which attributes a property to a statement, it simply blocks or bars the statement's assertion. Both Frege and John Searle have suggested (even claimed) in [2] and [9], respectively, that to deny statement A is simply to assert A's negation. But this reduction falsifies our experience of rejecting, or ruling out, assertions. And it fails to appreciate how statement negation gets its significance. Both assertion and denial are fundamental illocutionary acts, they can't be reduced to, or explained in terms of, some more fundamental acts. But once people perform acts of these kinds, they can reflect on the acts' criteria for success. For the assertion of a statement to be fully successful, the statement must fit the world, and the language user must have adequate grounds for accepting the statement. A fully successful denial must be of a statement which fails to fit the world, and the speaker must have adequate grounds for her denial.

Such reflection provides a rationale for characterizing a statement as true, or false. Denial is *prior* to negation. In negating a statement, one characterizes that statement as one that *deserves* to be denied — as one that fails to fit the world. In most circumstances, to deny a statement "comes to the same thing" as to assert that statement's negation. But to accept an act which characterizes a statement as false is quite different from ruling out an act of accepting the statement. It is worth noting here that a denial reflects no uncertainty about the statement's truth value. Searle has recognized in [9] an internal and an external negation associated with illocutionary acts. If I promise not to come, the negation is internal to the promise, while if (I say) I don't promise to come, the negation is external. It is unfortunate that Searle describes these both as negations, for the external negation is not one which characterizes anything. The external negation is part of an act of *declining* to perform an illocutionary act. I can also assert that it is *not* raining in Beijing or decline to assert that it *is* raining in Beijing. But the act of declining is not a denial, and it is not equivalent to a denial.

Once people learn to use language, and to make assertions and denials, they are in a position to *introduce* statement negation, and explain it in terms of denial with these principles:

$$\begin{array}{cc} \sim \text{Introduction} & \sim \text{Elimination} \\ \dashv A & \vdash \sim A \\ \hline \vdash \sim A & \dashv A \end{array}$$

If these principles characterize statement negation, then it makes no sense, it is irrational, to accept both a statement and its negation, just as it makes no sense to both assert and deny a single statement. For such negation, no special argument and no appeals to authority are needed to justify the principle of contradiction (or, perhaps, of non-contradiction). It is evidently correct. I think that the statement negation found in ordinary speech (which may not occur very frequently) does satisfy these principles.

4. A practical problem doesn't require a theoretical response

It sometimes (often?) happens that we find ourselves to be committed to make assertions and denials that are incoherent with one another. Finding this, we can be sure that some of our beliefs (and disbeliefs) are mistaken. We are probably entitled to be sure of this even when we don't find incoherence. But if we do find incoherence, and know that we are mistaken, we may not be able to fix the problem. We may not be able to determine which of our beliefs are mistaken, and which new beliefs we should adopt instead. What should we do, how should we proceed, when we know ourselves to be in error, but are unable to find, and fix, the source of the error?

Before discussing this question, I want to point out that we are dealing with a *practical problem*, not a logical one. It is a practical problem to determine what to do when we don't have enough money to pay our bills. It is a practical problem to determine what to do when we have made promises to different people which require us to be in two places at once. Everyone's life is filled with practical problems. If our problem is to respond to incoherence in our own beliefs, I think there are different strategies that can be employed, depending on the particulars of our case, or our situation. Although we hope to one day eliminate our erroneous beliefs, and replace them with more satisfactory ones, we have to live with the erroneous beliefs in the meantime, and employ them judiciously in dealing with matters that come up.

This practical problem provides no motivation for reconceiving matters so that the practical difficulties confronting us are in some sense "defined away". It is not an adequate response to insist that since incoherent beliefs are common, and perhaps inevitable, we should not regard them as a sign of error. When we realize that we are committed to accept inconsistent statements, or to perform incoherent assertive acts, this shows us that we are mistaken in some of our beliefs. It is actually a good thing to know about our mistakes, because then we are in a position where we can try to correct these mistakes. The problem of incoherence can't be solved by redefining implication or logical consequence so that incoherent beliefs and disbeliefs are made "respectable", or acceptable. Classical illocutionary logical theories, or illocutionary theories of classical logic, are both correct and normative. What is wanted, or needed, to deal with practical problems are something like engineering applications of objectively correct classical theories.

Theories of relevance logic and theories of paraconsistent logic can both be regarded as engineering applications of pure classical theories. Relevance theories can be seen as attempts to capture those rational commitments that people are likely to trace in carrying out real-life deductions. Such systems can be more useful than classical systems for computer implementations and simulations. For dealing with incoherent beliefs (or data bases) I think it might be useful to recognize a field, or class of theories, which are labeled *paracoherent logic*. In addition to genuine, or classical, logical consequence, we can also admit concepts of paraconsistent consequence, and devise deductive systems to be employed when we find ourselves with incoherent beliefs.

The need to devise these and other engineering applications of logical theory arises from issues concerning commitment rather than from issues involving truth and falsity. Not everything that follows logically from statements we accept has interest for us. We are rationally, but conditionally, committed to accept every statement that follows logically. For a variety of purposes, many of which are practical, we find it convenient to characterize a smaller class of what might be called *useful*, or *relevant*, commitments.

The statements in an inconsistent class aren't all true, they *can't* all be true. An inconsistent class *implies* every statement. But *people* can be *committed* to perform incoherent illocutionary acts. When this happens, those people are also committed, in an uninteresting sense, to accept every statement. What makes that commitment uninteresting is that everyone (every rational agent) has an over-riding commitment to achieve coherence. Incoherent beliefs (and disbeliefs) are a sign of error that needs correcting, they don't sanction making arbitrary assertions and denials. (It is slightly different with incoherent suppositions. These authorize us to suppose, as a conclusion, any statement at all. And this is sometimes a useful strategy for discharging a supposition.) For living with incoherence before we find a way to eliminate this incoherence, it can also be convenient to characterize useful, paracoherent, commitment relations leading from incoherent sets of assertions and denials to safe, or reliable, conclusions.

Illocutionary logic provides a highly general conceptual framework which accommodates, or incorporates, a great variety of more specific logical theories. Indeed, as I understand it, all deductive logical theories are illocutionary theories. Some are full theories containing two levels, an ontic and an epistemic level. The full theories have an explicitly first-person character, they are for a person to use to explore the commitments of her own assertive illocutionary acts. Other logical theories, standard logical theories, are proper parts of full theories, and have an apparently third-person character. However, even standard theories are concerned with language acts, which are intentional acts rather than some kind of impersonal objects.

The same basic kinds of assertive acts (assertions, denials, and suppositions, for example) are performed by speakers of different languages, and are characterized by the same commitments. The "central", and fundamental, illocutionary theories are classical. The generic versions of these theories are normative, for they explore and explain the commitments of language users whose beliefs and disbeliefs are correct, and who really know what they think they know.

5. Negation and pseudo-negation

It can't be correct to both accept a statement and rule that acceptance out of order. And if negation is related to denial as the introduction and elimination principles given earlier illustrate, then it can't be correct to accept both a statement and its negation. But it is certainly open to someone to dispense with our symbol for statement negation, to dispense, really, with our operation of statement negation. We can imagine someone simply refusing to employ this operation, claiming perhaps that our understanding of truth and falsity is mistaken, and we can further imagine that person making use of a "stand-in" for our negation.

To illustrate, and explore, this possibility, consider a very simple theory of propositional logic. This is a full theory of illocutionary logic, so it has both an ontic level and an epistemic level. The basic, ontic-level language contains atomic sentences, and compound sentences formed with these connectives:

$\wedge, \vee, *$

Disjunction and conjunction are understood in the normal way, but the asterisk is the symbol for *pseudo-negation*. The epistemic-level logical language is obtained from the ontic-level language by introducing the four illocutionary operators explained earlier.

The formal language of this theory constitutes the PN language. If we consider the first, or ontic, level theory for the PN language, we can come up with a system for constructing deductive derivations involving plain sentences of the language. To have some idea how to think of the asterisk, let us understand '$*A$' to mean that A is *difficult to accept, for objective reasons*. There is some particular feature of the statement A or of the situation that A describes or represents which makes it difficult for a person to accept A. All false statements are difficult to accept in the sense intended, and some true statements are difficult to accept. I won't make any proposals about what kind of feature this might be, for the whole idea of being difficult to accept is merely a heuristic device.

There are different deductive principles that might be adopted for '$*$', but it should be possible, for some sentences (statements) A to accept both A and $*A$ (to correctly perform both of these acts: $\vdash A$ and $\vdash *A$). Some pseudo-contradictions '[A & *A]' will be true (and acceptable), although the *law of pseudo-contradiction* '*[A & *A]' will be logically true, as will be the *law of pseudo-excluded middle*, '[A ∨ *A]'. A pseudo-contradiction doesn't imply every statement, and accepting a pseudo-contradiction doesn't commit a person to accept every statement. We can't specify the deductive principles for the asterisk, because I haven't said enough about what the asterisk means to fully determine what these should be. But we can certainly come up with an adequate set of principles.

Since the language contains no expression for "classical" negation, and no expressions for characterizing statements as false in the ordinary sense, it might seem that someone could learn the PN language, and "live" happily with pseudo-falsity and pseudo-negation, dispensing entirely with what I regard as genuine falsity and genuine negation. Someone might even think that this is the correct way to understand the language we actually do speak.

But this is a mistake. An ontic-level formal language doesn't fully represent the language acts we actually perform. An ontic-level language represents the materials we use in performing assertive illocutionary acts, but not those acts themselves. Once we add the illocutionary operators to the ontic-level PN language, which gives us the epistemic-level language, we gain the resources needed to characterize statements as "really" false.

Assertion and denial are not acts which represent or characterize expressions or objects in the world. They are acts of accepting and ruling out. Sometimes these acts are aimed at addressees, but this isn't essential. The two fundamental assertive acts are simple assertion and denial. These acts are the same no matter what language a person speaks, and they are not dispensable. We can accept a statement for fitting the world or rule out the assertion of a statement which fails to fit. Although the acts of assertion and denial are not acts which characterize statements, these illocutionary acts enable us to *introduce* expressions or operations which characterize statements as "really" true or "really" false. For we can't replace genuine denial by some sort of *pseudo-denial*.

We can't coherently both accept and deny a single statement. Most people wouldn't dream of doing this. We can, unfortunately, be *committed* to both accept and deny a single statement. If we are, then we are also committed, in an uninteresting sense, to assert every statement and to deny every statement. Most people wouldn't dream of asserting or denying arbitrary statements once they discover their own beliefs to be incoherent. For we have an overriding commitment to avoid incoherence in our beliefs, and to eliminate incoherence once it shows up. We only *exploit* incoherence in indirect arguments or proofs, where the fact that a supposed statement leads to incoherence gives us grounds for denying the statement that has been supposed.

Having incoherent beliefs is a sign of error. When this occurs, it presents a problem to be overcome. In a similar way, a genuinely contradictory statement is never true. Contradiction and incoherence may be fruitful, in stimulating us to avoid them, but this does not make them acceptable.

References

[1] Frege, G., 1966a, *Begriffsschrift*, chapter 1, in *Philosophical Writings of Gottlob Frege*, P. Geach & M. Black (eds.), Oxford: Basil Blackwell, 1-14.

[2] Frege, G., 1966b, "Negation", in *Philosophical Writings of Gottlob Frege*, P. Geach & M. Black (eds.), pp. 117-136.

[3] Kearns, J. T., 1997, "Propositional Logic of Supposition and Assertion", *Notre Dame Journal of Formal Logic*, vol. 38, pp. 325-349.

[4] Kearns, J. T., 2000, "An Illocutionary Logical Explanation of the Surprise Execution", *History and Philosophy of Logic*, vol. 20, pp. 195-214.

[5] Kearns, J. T., 2006, "Conditional Assertion, Denial, and Supposition as Illocutionary Acts", *Linguistics and Philosophy*, vol. 29, pp. 455-485.

[6] Kearns, J. T., 2007, "An Illocutionary Logical Explanation of the Liar Paradox", *History and Philosophy of Logic*, vol. 28, pp. 31-66.

[7] Kearns, J. T., 2010, "What is Natural about Natural Deduction", *The Logica Yearbook* 2009. Michal Peliš ed., London: College Publications, pp. 121-132.

[8] Priest, G., 2010, "Paradoxical Truth", *The Opinionator, Exclusive Online Commentary from the Times*, November 28.

[9] Searle, J. R., 1969, *Speech Acts: An Essay in the Philosophy of Language*. London: Cambridge University Press.

[10] Searle, J. R., 1985, *Expression and Meaning: Studies in the Theory of Speech Acts*. Cambridge: Cambridge University Press.

[11] Vanderveken, D. & and Searle, J. R., 1985, *Foundations of Illocutionary Logic*. Cambridge: Cambridge University Press.

John T. KEARNS
Department of Philosophy and Center for Cognitive Science
University at Buffalo, the State University of New York, USA

Approches Phénoménologiques de la Vérité Mathématique

FRÉDÉRIC PATRAS

1 Phénoménologie et logique

Les origines logiques de la phénoménologie sont bien connues : après la *Philosophie de l'arithmétique* de 1891, Husserl, confronté aux critiques de Frege et au développement, en mathématiques, de la méthode axiomatique de conception hilbertienne, a réorienté peu à peu sa pensée d'une approche psychologique en direction de préoccupations logiques. En témoignent entre autres ses différents articles sur la logique [35] et surtout les *Recherches Logiques* [34]. En dépit de différences de point de vue, la proximité d'intérêt entre ce qui allait devenir le courant phénoménologique et les approches plus techniques de la logique (Frege, Hilbert, Russell...) est alors flagrante.

Selon la formule célèbre de M. Dummett, souvent mise en exergue des études sur les rapports entre phénoménologie et philosophie analytique, Frege et Husserl peuvent ainsi "être comparés avec le Rhin et le Danube, qui naissent à proximité l'un de l'autre, font un bout de chemin parallèle, coulent ensuite dans des directions totalement différentes et finissent par déboucher dans des océans différents"[15, p. 44].

Cette divergence a souvent fait perdre de vue la proximité initiale, mais de multiples études ont, au cours des dernières années, conduit à réexaminer de façon constructive les rapports entre les deux philosophes et, plus généralement, ceux entre phénoménologie et philosophies de tradition analytique[1]. Cet intérêt pour la phénoménologie de la logique a des racines multiples, qui vont de la thèse d'A. Osborn[2] [39, p. 43] selon laquelle les critiques de Frege auraient joué un rôle déterminant dans l'évolution de la pensée du jeune Husserl et, par voie de conséquence, dans l'émergence de la phénoménologie, à l'essai de D. Føllesdal de 1958 [21]. Ainsi :

> "Now that the analytic tradition has become interested in its historical roots, the relations between Husserl's and Frege's views have

---

[1]On consultera en particulier [27, 9] et les travaux de J. Benoist sur la fertilisation croisée des deux courants philosophiques, dont, parmi d'autres, [2].

[2]Nous renvoyons à ce sujet aux études de R. Brisart [10] et J. English [17].

become a particularly important point of interest. Husserl has been
and can be considered via Frege's philosophy, but Frege can also be
considered in terms of Husserl's philosophy of logic; we may even
say that Husserl serves to give a deeper philosophical content to the
themes discussed by Frege." [27, Introduction]

2 L'approche heideggérienne

"Que le destin de ce qui au vingtième siècle a pu être nommé phénoménologie ait partie liée avec un remaniement du problème de la vérité, voilà ce dont personne ne pourra douter. Il suffira d'en prendre à témoin Heidegger, qui a su, sur ce point plus que sur aucun autre, mettre la rupture en scène, non sans en tirer à lui le profit." [1]

Plus encore que Husserl, Heidegger est en effet la référence phénoménologique obligée lorsqu'il s'agit de la vérité[3]. Celle-ci est indissociable, en tant que problème philosophique, de l'apophantique heideggérienne, enquête en direction du sens de l'être et des conditions de son dévoilement.

Ce sera le propos de cet article que d'essayer d'articuler le questionnement et les thèses heideggériennes sur la vérité aux théories plus classiques et surtout de mettre en évidence ce qu'ils peuvent apporter aux débats sur la vérité[4] et les fondements des mathématiques. Cette démarche s'inscrit dans un contexte plus général : bien que d'ampleur limitée relativement à ce qui s'est produit pour Husserl, plusieurs travaux ont été entrepris récemment pour engager un dialogue avec la logique à partir de l'œuvre de Heidegger avec, en particulier, les sessions du Séminaire des Archives Husserl de Paris en 1993-94 et 1994-95, travaux réunis dans le volume *Phénoménologie et logique* [13]. Le séminaire, orienté largement sur les rapports entre Husserl et la philosophie analytique, a cherché par ailleurs "à mettre concrètement à l'épreuve l'hypothèse d'un débat continu de Heidegger avec la logique et le versant "logique" de la phénoménologie husserlienne, ce qui procurait là encore un sol privilégié à la fois pour l'étude du phénoménologique chez Heidegger, de sa dette à l'égard de la percée husserlienne (intentionnalité, intuition catégoriale), et des enjeux du projet de "destruction de la logique", à la mesure d'un acheminement vers la parole". Nous reprenons ici rapidement, en suivant de très près les analyses de F. Dastur [14], quelques

---

[3]Signalons que les analyses de J. Benoist dans *Entre Acte et sens* [1] nuancent ce jugement. Les voies qu'il explore, dont les influences de Bolzano et Brentano sur la théorie husserlienne de la vérité, méritent de plein droit de figurer dans toute entreprise de relecture de la théorie phénoménologique de la vérité.

[4]Deux camps s'opposent dans la philosophie contemporaine à propos de la vérité. Un premier camp soutient une conception forte, "substantive", de la vérité : théorie de la correspondance, cohérentisme, pragmatisme... Un deuxième camp tient pour une conception faible : déflationnistes, minimalistes... Voir par exemple [16]

éléments permettant de situer les enjeux logiques généraux de l'œuvre heideggérienne et renvoyons à son remarquable article pour plus de détails.

L'intérêt de Heidegger pour la logique, inséparable de son intérêt pour la question métaphysique par excellence (celle de l'être, du sens de l'être, dans une tradition héritée d'Aristote et du projet même de sa *Métaphysique*), a motivé son intérêt pour les *Recherches Logiques* husserliennes et les recherches logiques et psychologiques de l'époque, puis pour la logique scolastique. Heidegger espérait alors d'une "logique du sens" la clarification du sens de l'être. Ces premières recherches, ces premiers essais indiquent déjà la direction générale vers laquelle sa pensée va s'orienter ultérieurement, en décalage avec les logiques qui nous sont aujourd'hui familières, mais en débat avec la phénoménologie husserlienne et l'histoire générale de la logique.

> "C'est en fait le même projet, celui du développement d'une *logique du sens orientée vers le jugement*, qui se distingue de la *logique au sens fort du terme, celui de la validité, orientée vers l'objet*, qui guide Heidegger dans son entreprise de Destruktion ou plutôt de kritischer Abbau der überlieferten Logik, de "déconstruction critique de la logique traditionnelle" [...]. Ce que Heidegger cherche à faire apparaître dans ce cours, c'est justement le statut métaphysique des principes métaphysiques initiaux de la logique, c'est-à-dire le caractère proprement philosophique de celle-ci qui, une fois accomplie la Destruktion de la théorie leibnizienne du jugement, c'est-à-dire la reconduction de celle-ci à ses fondements métaphysiques, peut être définie comme une "métaphysique de la vérité"." [14]

Le propos ultime de Heidegger est de comprendre l'essence de la logique, au-delà de sa présentation traditionnelle. L'enjeu n'est rien moins que celui d'une redéfinition des rapports entre logique et philosophie :

> "Mais alors il faut questionner radicalement l'essence de la vérité, ce qui ne veut pas dire que la métaphysique doive être reconduite à la logique ou le contraire, car ce qui est en question, ce n'est nullement la distribution des disciplines, mais ce sont plutôt les disciplines elles-mêmes qui sont le problème." [30]

Comme on le verra, l'un des intérêts de ce projet est de revenir aux questions de droit sur ce qu'*est* et *doit être* la vérité, là où la plupart des théories philosophiques de la vérité s'attachent plutôt à en donner une description normative –c'est l'un des enrichissements du débat que l'on peut attendre de la lecture de Heidegger, aussi délicate cette lecture soit-elle compte-tenu de l'idiosyncrasie de sa pensée. Heidegger, s'interrogeant sur la domination de la logique dans la pensée occidentale va remettre en question deux thèses de tradition aristotélicienne : Le jugement est le lieu de la vérité ; la vérité est adéquation de la pensée et de l'être. Il affirme ainsi la possibilité d'une vérité plus originaire, caractère de l'être même dont la vérité de la proposition et du jugement ne seraient que des caractères dérivés.

3 La vérité mathématique

Nous avons tenu à rappeler les grands principes de l'approche heideggérienne de la logique de façon à clarifier les termes mêmes des analyses qui vont suivre. Cette approche, on le devine, dans sa radicalité et son éloignement aux problèmes de la science moderne se prête mal à y être utilisée. Notre propos sera donc limité : il s'agira de montrer ce que la déconstruction heideggérienne de l'histoire et du fonctionnement de la logique peut apporter à son analyse et de mettre en évidence qu'elle peut compléter utilement les nombreuses théories, essentiellement descriptives, de la vérité en les mettant à l'épreuve d'une question de droit, celle de son essence.

Concrètement, il va donc s'agir non pas de réglementer ou de commenter les législations possibles de la vérité mathématique, mais plutôt de penser les conditions de possibilité théoriques et intentionnelles de la véracité des énoncés mathématiques, véracité dont l'exactitude formelle, pour être la plus remarquable et la plus significative, n'apparaît en fin de compte que comme l'une des modalités.

Il fallait choisir une voie d'approche pour penser les spécificités mathématiques de la vérité à la double lumière de la pratique scientifique contemporaine et de la phénoménologie. Fidèle, sinon aux intérêts spécifiques, du moins à l'esprit de la méthode employée par Dominique Janicaud au regard des textes heideggériens [36], cet article reproduira implicitement, et à sa manière, le mouvement d'idées du célèbre texte de Heidegger, *vom Wesen der Wahrheit* [29], partant de considérations historiques assez naïves sur la constitution de la vérité dans le corpus et les méthodes des mathématiques, pour en dégager peu à peu les traits les plus significatifs et les difficultés sous-jacentes.

4 La vérité est le réel

Qu'entend-on ordinairement par "vérité" ? Un tableau maniériste, une pierre précieuse sont dits vrais ou faux selon que leur apparence (celle d'un Tintoret, d'une émeraude) coïncide ou non avec leur être réel[5]. De la même

---

[5]Il pourrait être tentant de rapprocher la conception de la vérité comme "simplement" le réel d'approches minimalistes où la vérité d'une affirmation comme "la neige est blanche" se réduit à l'affirmation elle-même, la vérité ne lui ajoutant rien en substance. Ceci dit, il est assez difficile d'appréhender le sens de l'approche minimaliste dans un contexte mathématique, sauf peut-être à lui donner un sens frégéen, la vérité entendue chez lui comme référence des énoncés mathématiques vrais étant une notion extrêmement pauvre. En pratique, il est douteux qu'un tel rapprochement ait un sens compte-tenu des

manière, une égalité entre deux expressions algébriques sera dite vraie si sa forme (symbolique, égalitaire) est conforme à la nature même d'une égalité : exprimer l'identité réelle de ses deux constituants[6]. En ce sens, il est de faux Tintoret comme il est des identités algébriques fausses[7]. Cette conformité de l'apparence à la réalité est l'essence première de la vérité, mais repose sur une conception substantialiste, l'existence d'une réalité sous-jacente étant un présupposé métaphysique lourd, en particulier dans un contexte scientifique. Elle est d'emblée indissociable de ses moyens de contrôle : un test chimique ou spectroscopique pour l'émeraude, une vérification par le calcul pour l'identité algébrique.

Du point de vue des mathématiques modernes, cette notion de vérification se manifeste sous deux formes. Tout d'abord, une égalité ou un théorème sont vrais si on peut établir qu'ils sont tels –si on peut les démontrer. Le risque, du point de vue du problème épistémologique de la vérité, est de confondre ici fin et moyens, et de faire de la prouvabilité l'essence de la vérité, un peu comme si l'essence de l'émeraude était de satisfaire tel ou tel test spectrométrique. L'émeraude est véritablement émeraude si et seulement si elle satisfait à ces tests ; l'énoncé mathématique est vrai si et seulement si il est démontrable. Pourtant la vérité de l'émeraude et celle de l'énoncé excèdent, dans la perspective "réaliste"[8] sur la vérité, ces vérifications.

Dans le cas de l'émeraude, c'est évident. Sa présence, sa beauté, sa valeur, tout aussi réelles que sa composition chimique, sont indissociables du fait qu'il s'agit d'une vraie émeraude : de son "être-vrai" en un sens phénoménologique, qui englobe précisément ces différentes dimensions au-delà de la vérité conçue sur un mode purement assertorique. Dans le cas d'un énoncé mathématique, la situation est plus subtile, quoique sur le fond entièrement semblable. Le vrai, on l'a dit, est, selon cette première conception de la vérité, le réel. Ce Tintoret, cette émeraude sont vrais si, au-delà des apparences, on peut légitimement leur affecter leurs attributs

---

présupposés métaphysiques lourds qui grèvent la notion de réalité. Sur le minimalisme, voir par exemple [18] ou, pour un exposé de débats contemporains autour du thème de la vérité [16].

[6]Par "constituant" il faut entendre ici ce qui est désigné par chacun des termes.

[7]Ce type de phénomènes a été remarquablement analysé par Frege : "La vérité se dit de tableaux, de représentations, de propositions et de pensées. Il est remarquable que cette énumération réunisse des choses visibles et audibles et des choses qui ne sont pas perçues par les sens. C'est là l'indication d'un déplacement de sens, etc" [24]. Voir aussi "Sens et dénotation" [25]. Le point de vue qu'il adopte est toutefois différent de celui que nous suivons dans ce premier moment de l'analyse. Nous reviendrons ultérieurement sur la conception frégéenne de la vérité.

[8]Les guillemets autour de réaliste indiquent la volonté de désigner ici simplement la première thèse, la vérité est le réel, la notion de réel étant entendue en un sens naïf –sans entrer donc dans de quelconques débats sur le réalisme comme courant philosophique.

naturels –une charge émotive, historique ou esthétique, par exemple, qui participe, tout autant que l'exactitude de l'attribution d'un tableau ou la composition chimique d'une pierre, à leur essence d'objet d'art ou de pierre précieuse.

De façon plus indirecte, un théorème est vrai lorsque, au-delà de l'existence d'une preuve, il peut être légitimement chargé d'un contenu théorique et s'insérer dans un horizon mathématique[9]. Les mots sont importants ici, un mathématicien n'appellera pas "théorème" (ni même "lemme", "corollaire"...) un énoncé arbitraire mais uniquement un énoncé prouvable où se cristallisent des enjeux conceptuels : solution d'un problème jugé important, résultat stratégique dans tel ou tel domaine... La pensée mathématique s'organise autour de schémas dont la vérité n'est qu'une composante, essentielle mais en fin de compte subordonnée à des enjeux sémantiques, stratégiques et opératoires.

De fait, la vérité mathématique est une notion assez pauvre si on l'assimile à la prouvabilité. Un ordinateur, à supposer même qu'il puisse reconstruire les mathématiques sur le fondement de processus déductifs mécaniques, ne saurait pas pour autant sélectionner les énoncés théoriquement intéressants dans la masse des énoncés exacts. Tout mathématicien ayant utilisé des logiciels de calcul formel a ainsi été confronté à la tâche de dégager des formes algébriques typiques à partir de listings informatiques où les symboles et identités se perdent dans l'anonymat des formules. Ce type de phénomène d'émergence de structures, où le mathématicien détecte des phénomènes typiques pour leur donner une forme générale, abstraite, est sans doute l'un des phénomènes les plus remarquables de la pensée mathématique, mais son articulation au formalisme est difficile à penser en dehors de l'expérience, c'est-à-dire du phénomène même du travail mathématique.

Plus radicalement, dans la pratique mathématique quotidienne, certains mathématiciens ont pu soutenir que le pire ennemi du vrai n'était pas le faux mais bien l'insignifiant –et ce non sans légitimité, puisque le faux porte en lui-même les conditions de possibilité de son dépassement, et donc d'un accroissement de la connaissance théorique, là où l'insignifiant est sans portée aucune. L'identification du vrai au démontrable, quelle que soit sa légitimité

---

[9]L'horizon mathématique désigne ici l'ensemble des éléments qui participent à sa compréhension et son utilisation, de façon très générale (y compris, par exemple, des conjectures au statut incertain, des analogies encore imprécises...). Il s'agit donc d'une occurrence particulière de la notion philosophique d'horizon telle qu'on la trouve décrite, par exemple, dans [33] : "Une chose est nécessairement donnée sous de simples "*modes d'apparaître*", on y trouve donc nécessairement un *noyau* constitué par ce qui est "réellement figuré" et, autour de ce noyau, au point de vue de l'appréhension, tout un horizon de "*co-données*" dénuées du caractère authentique de données et toute une zone plus ou moins vague d'indétermination."

factuelle, occulte cette possibilité de penser le mode de fonctionnement de la pensée mathématique authentique[10], entraînant un rétrécissement dramatique des ambitions de toute philosophie mathématique. Ces remarques ont d'ailleurs une portée concrète en informatique théorique où les tentatives d'automatisation de la construction de preuve achoppent toujours à l'irréductibilité des mathématiques effectives à leurs modèles formels.

La perspective "réaliste" est par ailleurs indissociable du problème de l'ontologie des énoncés mathématiques et est une des modalités d'appréhension du problème du "platonisme des mathématiciens". Les thèses du platonisme gödélien modéré en arithmétique et la thèse d'A. Connes selon laquelle il existerait, pour l'arithmétique, une "réalité archaïque" (celle des résultats vrais, mais indémontrables dans un système formel donné) entrent ainsi dans cette perspective épistémologique[11]. La charge métaphysique et ontologique qui est ainsi associée à la conception "réaliste" de la vérité est l'une des difficultés qui pèsent sur sa justification philosophique, à l'image des débats sur le platonisme.

5 La validation empirique

La seconde forme de l'idée de vérification dans les mathématiques modernes, quoique moins prégnante, a ce mérite de rendre partiellement compte de leur applicabilité. Un énoncé mathématique –comme une égalité numérique– peut être directement vérifiable. Il peut également porter sur des quantités ou des êtres génériques : "tout nombre réel positif admet une racine carrée" ; "tout corps fini est commutatif" (théorème de Wedderburn). Dans un tel énoncé, on peut substituer à l'élément générique (nombre réel positif, corps fini) une valeur spécifique ou un modèle (2, le corps à 9 éléments) et vérifier l'énoncé générique une fois la substitution opérée. Cette vérification peut d'ailleurs prendre diverses formes : dans le cas du théorème de Wedderburn, on pourra ainsi chercher par exemple à construire et classifier les corps à 9 éléments à isomorphisme près puis vérifier que l'unique solution est un corps commutatif.

Dans certaines situations, quelques substitutions bien choisies peuvent permettre d'établir l'énoncé général[12]. Très souvent, les conjectures mathématiques sont formulées par ce procédé : quelques calculs sur des cas

---

[10]La possibilité donc de penser les modalités et ressorts de la pratique mathématique (opératoire, conceptuelle, eidétique...).

[11]Voir par exemple [41].

[12]Exemple classique : l'égalité de deux polynômes définis par des conditions implicites ou des identités complexes (et que l'on ne peut donc pas comparer immédiatement en comparant deux à deux leurs coefficients) peut être établie en comparant les évaluations de ces deux polynômes en un nombre fini de valeurs de la variable, ce nombre ne dépendant que du degré des deux polynômes.

particuliers, parfois très simples, suffisent à indiquer, voire à convaincre de la validité probable de conjectures très générales. Le théorème de Fermat ou les conjectures de Weil en sont deux exemples célèbres. Dans le cas de ces dernières, d'un énoncé général complexe[13], la précision de l'énoncé constrate singulièrement avec le caractère relativement élémentaire des exemples ayant servi à établir et étayer la conjecture (le cas des courbes, c'est-à-dire des variétés de dimension 1). Sauf à refuser aux conjectures la place centrale qu'elles doivent légitimement occuper en mathématiques, on ne peut que constater qu'en elles le rapport du mathématicien à la vérité ne peut être pensé sur le mode de l'exactitude. De fait, l'expérience va jusqu'à laisser penser que même des approches comme la logique modale et ses variantes probabilistes sont incapables de rendre pleinement compte de ce rapport très particulier du mathématicien à la vérité qui se fait jour dans la vraisemblance d'une conjecture. La signification des énoncés (y compris faux ou conjecturels), bien qu'ayant naturellement sa place dans une conception étendue de la logique, s'avère ainsi difficile à quantifier ou à modéliser.

Pour en revenir au schéma de substitution général, il est possible de lui donner une signification très concrète, classiquement associée à l'opposition de style entre mathématiques archimédiennes et platoniciennes. Lorsqu'une théorie mathématique est un modèle formel pour tel ou tel ordre de phénomènes (physique, chimique, biologique...), la vérité (comprise ici au sens de la prouvabilité, de l'exactitude formelle) d'un énoncé est indissociable, en termes cognitifs et scientifiques, de la vérité (comprise au sens de conformité au réel expérimental) de la propriété phénoménale correspondante[14]. Toutes les limites que l'on peut poser à ce principe de permanence de la validité des lois mathématiques par substitution et passage à des modèles physiques (limites de validité des modèles, etc.) n'altèrent en rien la force et l'universalité du principe, faute duquel les mathématiques ne seraient qu'un jeu de langage parmi d'autres, dépourvu de toute exemplarité du point de vue des sciences de la nature[15].

---

[13]L'énoncé porte sur les séries génératrices associées au nombre de points de variétés algébriques sur des corps finis.

[14]Phénomène que la thèse du holisme épistémologique (Quine) théorise et radicalise.

[15]Cette seconde approche à la vérité et ces questions peuvent être rapprochées des thèses pragmatistes (Ch. S. Pierce, W. James, J. Dewey pour sa conception "classique"; le pragmatisme connaît une renaissance depuis les années 1970 avec des philosophes comme R. Rorty ou H. Putnam) qui voient dans le caractère pratique et l'efficacité prédicative les critères de choix des schémas conceptuels. Le problème de la faillibilité des énoncés mathématiques lorsqu'il s'agit de décrire la réalité phénoménale (on peut penser aux débats sur le caractère non euclidien de l'espace) prend naturellement place dans ce cadre. Les situations où la vérité des énoncés mathématiques est suggérée par des "expérimentations" numériques et théoriques forment un champ d'étude intéressant de

En ce sens, la vérité mathématique, quelque signification qu'on lui accorde, n'est jamais purement formelle. Il y a, dans tout résultat mathématique théorique, une dimension de facticité et d'applicabilité qui excède sa validité formelle et renvoie à l'intelligibilité globale des phénomènes. L'idée parcourt, de manière problématique, toute l'histoire des mathématiques et leur philosophie, l'époque récente se caractérisant peut-être par un malaise affirmé à l'égard de ce type de questions par trop "métaphysiques".

6 Adæquatio intellectus ad rem

Si la vérité se caractérise d'abord comme conformité de l'apparence à la réalité, à l'être réel de l'objet, sous un mode plus ordinaire elle ne nous apparaît pas directement comme vérité de l'objet, de la "chose" examinée, mais plutôt comme vérité du jugement sur l'objet (l'émeraude) ou sur la "chose" examinée (une propriété d'objet mathématique, par exemple). Comme souvent, la nuance entre ces deux modes d'être du vrai prend en mathématiques une dimension spécifique, qui met en relief une de leurs caractéristiques : elles n'ont d'existence intersubjective que comme langage, sans que les "objets" auxquels elles renvoient n'aient de réalité matérielle. Pour cette raison, il y est sans doute plus difficile qu'ailleurs de concevoir une vérité qui ne se jouerait pas d'emblée dans l'élément du langage et de l'énonciation.

On reconnaît dans cette caractérisation de la vérité une modalité de la thèse aristotélicienne classique selon laquelle la vérité est adéquation de la pensée et de l'être[16]. La formule thomiste qui énonce classiquement cette dimension noétique de la vérité (*veritas est adæquatio intellectus et rei* et la variante selon laquelle un jugement est vrai s'il se conforme à la réalité extérieure) a le mérite de retenir l'attention sur deux question fondamentales pour le problème de la vérité : statut de l'objet ; nature du rapport d'adéquation entre langage et objets et problèmes de priorité et de constitution afférents : la pensée et le langage qui l'exprime se règlent-t-il spontanément sur des domaines d'objets mathématiques idéaux (ceux de la géométrie, par exemple), ou jouent-t-ils un rôle constitutif (les domaines d'objets mathématiques étant inséparables du discours théorique qui les porte à l'existence) ?

La vulgate platonicienne, qui consacre l'objectivité à part entière des objets mathématiques, si elle résout superficiellement le problème et est

---

ce point de vue, puisqu'elles placent les problèmes de validation des énoncés à l'intérieur même du domaine scientifique, qui devient son propre domaine d'empirie.

[16] Il s'agit là de la conception philosophique la plus commune de la vérité avec, comme exposants modernes, Moore, Russell ou Tarski ; elle a été largement remise en cause au cours du vingtième siècle.

communément adoptée comme pis-aller, ne va pas sans difficultés, comparables dans leur complexité à celles qu'elle entend résoudre. Pour aller au-delà, il faut remarquer que l'adéquation de la pensée mathématique à son objet peut, dans une certaine mesure, se mesurer à la même aune que celle de l'apparence de l'objet à son être réel : l'adéquation se fait lorsqu'elle est réelle, lorsque la pensée s'accorde aux propriétés réelles de l'objet. Mais, là encore, insistons-y : les problèmes de l'objectivité et de la réalité mathématique se doivent d'être d'abord réglés pour que cette adéquation soit intelligible : il semble bien qu'il ne soit pas d'entente possible de la vérité sans un détour par l'ontologie[17].

L'écart entre cette théorie de la vérité (dite théorie de la correspondance) et la thèse d'un simple agrément au réel tient entre autres à son insistance sur le jugement et les propositions. Cet écart permet de dégager un espace de réflexion théorique plus facilement compatible aux approches syntaxiques et formelles aux théories mathématiques qu'un "réalisme" naïf, tout en s'accompagnant de thèses ontologiques moins fortes.

7 Vérité et formalisme

C'est à Frege que l'on doit d'avoir posé avec la plus grande clarté conceptuelle la question des rapports entre objectivité, formalisme et vérité en mathématiques. De toute son œuvre, l'une des conclusions les plus frappantes pour un mathématicien est sans nul doute le constat d'une immense pauvreté du concept *formel* de vérité[18]. Revenons sur un exemple classique et significatif pour en illustrer les ressorts : le théorème de Wedderburn, qui s'énonce "tout corps fini est commutatif". Pour comprendre l'essence de la vérité (formelle, donc en s'interdisant tout recours à une signification qui déborderait le formalisme) associée à cet énoncé, il faut en comprendre la référence : ce qui est visé par l'énoncé en tant que tel. L'interprétation naïve consiste à faire de ce théorème l'énoncé d'une propriété d'un objet (tel corps fini, comme le corps à 9 éléments) ou d'une classe d'objets (les corps finis). La vérité de l'énoncé serait alors indissociable de la vérité de certaines propriétés remarquables des corps finis, cette dernière vérité étant

---

[17] Sur ce thème, on pourra consulter [37]. J.N. Mohanty s'y intéresse à la "logique de la vérité" husserlienne et ses rapports à l'esthétique transcendantale.

[18] Cette pauvreté est illustrée par la formule célèbre des *Recherches logiques* [24] selon laquelle "Il y a tout lieu de penser que nous ne pouvons pas reconnaître qu'une chose a une certaine propriété sans en même temps estimer vraie la pensée que cette chose a cette propriété. Ainsi à toute propriété d'une chose est liée une propriété d'une pensée, à savoir celle d'être vraie. Il vaut aussi de remarquer que la proposition "je sens une odeur de violette" a même contenu que la proposition "il est vrai que je sens une odeur de violette". Il semblerait que rien n'est ajouté à la pensée quand je lui attribue la propriété d'être vraie." On reconnaît là une position typique du courant minimaliste auquel Frege est parfois rattaché.

conçue au sens d'adéquation. C'est l'interprétation classique, ontologisante, dont Frege se démarque radicalement[19].

Bien entendu, c'est la fonction du théorème que d'énoncer une propriété des corps finis, mais il faut bien se garder de confondre les objets *sur lesquels* porte l'énoncé de la référence de l'énoncé, qui est d'une toute autre nature. De fait, l'idée d'une vérité qui se jouerait directement au niveau de l'objet et non au niveau de l'énoncé n'a pas sa place dans le système frégéen. Il est d'ailleurs douteux que, du point de vue formel, les objets puissent avoir une existence autonome : ils coexistent avec les fonctions propositionnelles où ils sont susceptibles d'intervenir et, à supposer que l'on puisse encore parler d'objet dans ce contexte sous une autre forme que dérivée, n'ont pas de primauté ontologique sur les énoncés qui les définissent ou décrivent leurs propriétés[20].

Non, dans la perspective frégéenne, la référence du théorème de Wedderburn (ce qui, dans l'énoncé, est visé) n'est pas l'objet (générique, et problématique dans sa généralité) "corps fini", mais bien la valeur de vérité de l'énoncé. Comme une valeur de vérité est par essence de nature booléenne –la notion de valeur de vérité est codée mathématiquement par une fonction à valeurs 0 ou 1–, c'est tout le problème de la vérité qui semble se dissoudre dans une vaste tautologie, un abîme pour la pensée réflexive confrontée à la mécanisation de la logique et à l'idée (très impressionnante pour un mathématicien débutant) que tous les énoncés mathématiques vrais sont logiquement équivalents et donc fondamentalement indiscernables du seul point de vue de leur être vrai[21].

Ces thèses ont pour corollaire celles de Wittgenstein, selon lesquelles : "Les mathématiques sont une méthode logique ; les énoncés des

---

[19]B. Bégout souligne que "l'indifférence de Frege aux questions épistémogiques a conduit bon nombre de philosophes de la tradition analytique à se détacher progressivement de lui [...]. Ces accusations sont encore plus vives chez les partisans de la philosophie de l'esprit qui tentent de renouer les liens entre philosophie et psychologie et donc de dépasser l'antipsychologisme de Frege" [4]. Le renouveau d'une philosophie de l'esprit milite évidemment pour une meilleure prise en compte de la tradition phénoménologique.

[20]Dans la logique frégéenne, l'objet est d'abord caractérisé par son opposition aux concepts ou fonctions, par essence insaturés : en écriture symbolique, dans l' expression $f(A)$, $f(-)$ est la fonction (insaturée) et A, qui intervient comme argument, l'objet. Pour autant, comme Frege le remarque très vite, et conformément à la conception contemporaine de la dualité, si l'on définit $A(f)$ par $A(f) := f(A)$, objet et fonction sont amenés à échanger leurs rôles. Lorsque f est une forme linéaire et A un vecteur, la dualité dont il s'agit est la dualité ordinaire en géométrie affine. Dans le cas où A est un objet mathématique (le corps à 9 éléments) et f une fonction propositionnelle (l'énoncé du théorème de Wedderburn), on voit bien comment ce phénomène de dualité tend à rendre évanescente la notion même d'objet.

[21]Au sens ici selon lequel Frege affirme que "Découvrir des vérités est la tâche de toutes les sciences, mais c'est à la logique qu'il appartient de connaître les lois de l'être vrai". [23]

mathématiques n'expriment aucunes pensées"[44, (6.2;6.21)]. Un non-mathématicien peut peut-être se satisfaire du constat d'une tautologie des mathématiques qui en découle, mais un mathématicien doit rester nécessairement assez perplexe sur sa véritable portée. La vérité est une notion suffisamment riche et complexe pour que son sort ne puisse être décidé trop rapidement.

Dans la veine même de la pensée frégéenne[22], deux voies s'offrent pour dépasser ce constat de pauvreté formelle. La distinction *Sinn/Bedeutung*, sens/référence, en est une, la plus communément suivie [23]. Si la référence du théorème de Wedderburn est sa valeur de vérité, sa signification en prend en charge le mode de donation : par exemple, tout ce qui est généralement donné comme acquis dans un cours de mathématiques lorsque le théorème est formulé pour la première fois, comme l'existence de corps finis de cardinaux les nombres premiers, etc. Si elle ne remet pas en cause une conception booléenne de la vérité, cette approche a le mérite de dégager un terrain où penser les ressorts de la pratique mathématique.

Une deuxième voie est plus intéressante encore du point de vue du problème spécifique de la vérité. Tout le projet frégéen tel qu'il est énoncé dans les *Fondements de l'arithmétique* [22] vise à fonder les mathématiques et, plus précisément, l'arithmétique, sur les lois pures de la pensée. En ce sens-là, et quelles qu'aient pu être les interprétations ultérieures de Frege, dès lors que l'on interprète cette idée de lois pures de la pensée comme une forme antéprédicative de nécessité, son projet continue de pouvoir être rattaché au pacte apophantique aristotélicien, selon lequel la vérité mathématique est indissociable de l'idée de nécessité, et cela y compris lorsque le système déductif est reconduit à un système d'axiomes[23]. Le débat entre Frege et Hilbert sur l'étendue des pouvoirs de la méthode axiomatique[24] [26] devait précisément mettre en évidence ce qui aujourd'hui semble souvent archaïque dans l'œuvre frégéenne, mais pourrait bien témoigner de l'un des impensés majeurs de l'épistémologie mathématique

---

[22]Le constat d'un caractère tautologique des mathématiques est profondément étranger à la pensée du "platonicien" Frege : voir par exemple les *Recherches logiques* [24].

[23]Dans la pensée aristotélicienne, un axiome est une thèse dont la vérité est première : C'est une hypothèse ayant un caractère de nécessité. L'affranchissement de l' idée d' axiomatique de celle de nécessité est contemporain des travaux de Frege, mais ne lui est pas dû : si ses sources sont multiples, il est généralement attribué à Hilbert, dont les *Fondements de la géométrie* [32] marquèrent les débuts de la conception moderne de l'axiomatique.

[24]Sur les liens entre mathématiques et réalité chez Hilbert, et sur les débats avec ses contemporains, voir [5].

du vingtième siècle[25]. Frege reste ainsi éminemment actuel[26], et les progrès de l'épistémologie contemporaine doivent passer de toute nécessité par une confrontation avec la dimension la plus radicale –et sans doute la plus oubliée– de son projet : cette volonté de rendre compte des "lois pures de la pensée", de la logique en tant qu'elle n'est pas simplement formelle mais organiquement condition de possibilité de toute pensée rationnelle.

8 Le structuralisme et la thèse de la robustesse ontologique

L'un des grands phénomènes de la seconde moitié du vingtième siècle mathématique a été l'émergence, la domination puis le déclin de la pensée structurale[27]. Rappelons brièvement que, pour les grands mathématiciens des débuts du vingtième siècle, et quelles qu'aient pu être les difficultés soulevées par les paradoxes de la théorie des ensembles, la dimension empirique du savoir mathématique, aussi problématique puisse-t-elle être en droit, allait de soi en pratique. Les figures de proue de l'époque comme Poincaré et Hilbert, mais également les Hadamard, Elie Cartan, et al., puis les élèves de l'école hilbertienne (H. Weyl, J. von Neumann), étaient indissociablement mathématiciens, mécaniciens, logiciens et physiciens. Les conceptions de la vérité et de la pratique mathématique qui imprègnent les réflexions de Poincaré, Hilbert ou Weyl sont incompréhensibles si l'on perd de vue l'enracinement de leur travail mathématique dans une conception globale de la pensée scientifique, nourrie du dialogue avec les sciences physiques et l'épistémologie. Un beau discours[28] du géomètre Felix Klein, prononcé à Vienne en 1894 en l'honneur de Riemann illustre bien les préoccupations et l'état d'esprit de l'époque. Klein, qui connaît les motivations physiques de Riemann, y prend soin de distinguer la question de l'origine des théories de la connexion logique des axiomes, les moments génétiques et formels d'une théorie mathématique ne devant pas être opposés mais considérés comme complémentaires. Le discours précise cette

---

[25]Il faut exclure de cette remarque les travaux husserliens à partir des *Recherches logiques* et tout particulièrement les idées d'ontologie formelle et de logique transcendantale.

[26]Rappelons à ce propos l'existence et l'actualité d'une "lecture frégéenne de la phénoménologie" dont l'objectif est de faire à l'aide de Frege et pour l'interprétation de la phénoménologie ce qui a été fait pour la philosophie analytique. Cette lecture met surtout l'accent sur les rapports avec Husserl. Outre les travaux déjà mentionnés, il convient de citer ici tout particulièrement [19, 20].

[27]Le structuralisme dont il est question ici est le structuralisme classique des mathématiciens, celui de Bourbaki tel qu'on le trouve étudié dans [12, 40]. Le structuralisme moderne s'intéresse aux phénomènes ontologiques attenants, mais d'un point de vue différent [43].

[28]Disponible dans la traduction française des *Œuvres Complètes* de Riemann [42].

articulation :

> "Il n'est pas indifférent, dans la recherche et la découverte des lois mathématiques, d'attribuer ou non aux symboles avec lesquels on opère une signification déterminée. En effet, la présentation concrète nous fournit la liaison des idées qui doit nous conduire en avant [...]. Les résultats qui dérivent des recherches des mathématiques pures sont pourtant au-dessus de tout type de particularisation. C'est un schéma général logique, système dont le contenu particulier n'est pas indifférent, ce contenu pouvant être choisi de manières diverses."

Généralité et validité d'un schéma logique d'une part, rôle et non-indifférence de la signification d'autre part. Klein refuse pourtant une dichotomie du symbolique et de la pensée intuitive. Hilbert ne s'exprimera pas différemment au début de ses Fondements de la géométrie [32] ni dans son œuvre postérieure.

Le structuralisme mathématique, qui est né dans les années 40 et s'est imposé rapidement, s'est accompagné d'un repli autarcique des mathématiques, témoignant de l'émergence d'un nouveau type de pratique des mathématiques et d'une nouvelle génération de mathématiciens. Les problèmes internes à la discipline (comme l'étude des phénomènes arithmétiques) prirent le pas sur ceux soulevés par les sciences de la nature (comme ceux liés à l'émergence de structures mathématiques complexes en mécanique quantique). Bien entendu, les premiers fondateurs du structuralisme mathématique (A. Weil et J. Dieudonné en particulier [3]) n'avaient, de par leur formation et leurs intérêts, ni les moyens théoriques ni la volonté de thématiser leur rapport conflictuel aux sciences de la nature ou a fortiori au problème général de la vérité mathématique[29], mais les quelques témoignages qui nous sont parvenus sont significatifs et témoignent d'un changement profond :

> "Le mode de raisonnement par enchaînement de syllogismes n'est qu'un mécanisme transformateur [...]. C'est la forme extérieure que le mathématicien donne à sa pensée, le véhicule qui la rend assimilable à d'autres et, pour tout dire, le langage propre aux mathématiques [...]. Le formalisme logique –nous insistons sur ce point– n'est qu'une face, la moins intéressante, de la méthode axiomatique. Ce que celle-ci se propose pour but essentiel, c'est précisément ce que le formalisme

---

[29]Dans le texte manifeste du structuralisme, les Éléments de mathématiques de Bourbaki, dont la parution commence dans l'après-guerre, les choix épistémologiques effectués le sont a minima : pour l'essentiel, Bourbaki adopte la méthode axiomatique sous la forme dérivée des travaux de Russell-Whitehead via Zermelo-Fraenkel et al. Par ailleurs, Bourbaki affirme résolument et régulièrement son désintérêt pour les problèmes de fondements et les débats qui ne seraient pas internes à la discipline.

logique, à lui seul, est incapable de fournir, l'intelligibilité profonde des mathématiques." [8]

Pour la pensée structurale, l'essence de la vérité mathématique a donc une dimension qualitative et, en tant qu'elle est organisée en un système, doit être reconduite, outre à un système d'axiomes, à une architecture globale du corpus mathématique. L'idée est loin d'être inintéressante et aurait pu conduire à un dépassement du point de vue formaliste si elle avait été développée pour elle-même, et si les problèmes épistémologiques attenants avaient été abordés avec la rigueur nécessaire. Occasion manquée, à laquelle les morts prématurées de Cavaillès, Herbrand et Lautmann, qui auraient pu réconcilier les mathématiques françaises de l'après-guerre avec leur épistémologie, ne sont sans doute pas indifférentes.

Sur le plan mathématique, le programme structuraliste, s'il s'est parfois avéré très fécond, ne va pas sans poser divers problèmes qui en limitent la validité ou, du moins, le champ d'application. La recherche de la plus grande généralité, inséparable de la méthode axiomatique, mais poussée à son paroxysme dans l'écriture mathématique structuraliste, s'accompagne nécessairement d'une perte de vue des phénomènes individuels. L'objet mathématique concret se dissout dans la structure et la théorie qui le décrivent : "Dans cette nouvelle conception [le structuralisme], les structures mathématiques deviennent, à proprement parler, les *seuls "objets"* de la mathématique"[8]. Dans cette perspective, toute vérité intéressante serait générale, universelle. Or, l'expérience du travail mathématique enseigne que l'attention aux spécificités des objets (au sens usuel du mot) est une des composantes essentielles du travail et du progrès mathématique – tout aussi indispensable que l'aspiration à la plus grande généralité possible des énoncés.

L'autre objection à la pensée structuraliste tient à son incapacité à penser l'enracinement phénoménal des mathématiques. Bourbaki va jusqu'à parler, dans les années 50, de "contact fortuit entre les mathématiques et la physique". Là encore, au-delà de goûts personnels chez les mathématiciens de l'époque, qu'il ne faut pas sous-estimer, les raisons de cette reconduction des mathématiques et de leur mode de véracité à leur structure interne, axiomatique-formelle, tiennent à une incapacité à valoriser la notion d'individu (mathématique), de modèle, de réalisation d'un système axiomatique. La particularisation des axiomes, inhérente à tout système physique, s'accompagne, du point de vue structural, d'une dévalorisation des vérités correspondantes (des énoncés typiques du système physique), d'où un certain mépris pour la physique mathématique. Celui-ci a joué un rôle néfaste au sein des mathématiques françaises débordées, dans les années 80, par la créativité des mathématiques russes qui s'étaient engagées sur des voies radicalement différentes.

Pour ce qui est de la thèse revendiquée de réduction de l'ontologie au phénomène des structures, elle doit être relativisée et l'on trouve, à l'intérieur même du corpus bourbakiste, des éléments pour la nuancer. Le bon sens mathématique des membres de Bourbaki les a sans doute empêchés de prendre trop au sérieux les revendications méthodologiques de l' "Architecture des mathématiques" [8]. Ainsi, allant en direction d'une dimension intuitive et pragmatique de la connaissance mathématique (d'un pragmatisme original, fondé largement dans la fréquentation des idéalités!) :

> "*La faculté de donner des contenus multiples aux mots ou notions premières d'une théorie est une source importante d'enrichissement de l'intuition du mathématicien, qui n'est pas nécessairement de nature spatiale ou sensible comme on le croit parfois, mais qui est plutôt une certaine connaissance du comportement des êtres mathématiques.*" [6]

D'autres passages encore font partiellement écho à la conception pragmatiste de Pierce selon lequel la vérité est la concordance d'un énoncé abstrait avec la limite idéale vers laquelle une quête sans fin tendrait à conduire la croyance scientifique :

> "*Le mathématicien qui désire s'assurer de la "rigueur" d'une démonstration ne recourt guère à une formalisation complète [...], il se contente en général d'amener l'exposé à un point où son expérience et son flair de mathématicien lui enseignent que la traduction en langage formalisé ne serait plus qu'un exercice de patience (sans doute fort pénible). Si des doutes viennent à s'élever, le redressement se fait invariablement, tôt ou tard, par la rédaction de textes se rapprochant de plus en plus d'un texte formalisé jusqu'à ce que, de l'avis général des mathématiciens, il soit devenu superflu de pousser ce travail plus loin.*" [6]

Toutes ces nuances suggèrent que l'épistémologie implicite du structuralisme mathématique classique (par opposition à ses variantes récentes, qui ont un rapport beaucoup plus critique à l'ontologie) repose, dans les faits, sur la croyance de fond en une robustesse ontologique des vérités mathématiques. La conclusion de l'Introduction à la *Théorie des Ensembles* [6] en témoigne amplement :

> "*Nous croyons que la mathématique est destinée à survivre, et qu'on ne verra jamais les parties essentielles de ce majestueux édifice s'écrouler du fait d'une contradiction soudain manifestée ; mais nous ne prétendons pas que cette opinion repose sur autre chose que l'expérience. C'est peu diront certains. Mais voilà vingt-cinq siècles*

que les mathématiciens ont l'habitude de corriger leurs erreurs et d'en voir leur science enrichie, non appauvrie ; cela leur donne le droit d'envisager l'avenir avec sérénité."

9 L'essence de la vérité

Les remarques de Frege sur les nombres généralisés (comme les imaginaires) dans les derniers paragraphes de ses *Fondements de l'arithmétique* [22], archaïsantes du point de vue des mathématiques modernes, indiquent, tout comme son refus de conclure de la non-contradiction formelle d'un système d'axiomes à l'existence des objets correspondants, encore une difficulté : lorsque la vérification de la non-contradiction d'un tel système devient un argument ontologique permettant de conclure à l'existence des objets et de la théorie correspondants, on confondrait un critère nécessaire et ses corrélats.

S'il paraît légitime de penser aujourd'hui encore que le problème ontologique de l'existence des objets mathématiques ne se réduit pas à la non-contradiction, ce n'est cependant pas au nom d'une naïveté de mauvais aloi. Nombre de mathématiciens sont ainsi convaincus, de par leur expérience même, de l'existence d'un monde d'objets mathématiques existant en soi. Comme le suggèrent les analyses de Bourbaki, la vérité mathématique est indubitablement beaucoup plus "robuste" que ne le laisserait penser une approche formalisante, et tout mathématicien qui a suffisamment réfléchi à la nature de son activité sait bien que ses idées ne dépendent qu'assez faiblement des choix de systèmes d'axiomes sous-jacents, comme de ceux effectués en théorie des ensembles. C'est que, pour l'essentiel, le travail mathématique est de nature locale : pour résoudre telle ou telle difficulté, il n'est qu'exceptionnellement nécessaire de revenir aux définitions et aux axiomes originels. La pensée s'oriente dans le problème, guidée par des structures (combinatoires, algébriques, géométriques...) sous-jacentes, qu'elle découvre et ordonne en une démonstration ou une théorie au fur et à mesure de sa progression vers la solution. La plupart des mathématiciens sont ainsi conduits à adopter spontanément une certaine forme de réalisme ontologique[30]. Pour peu que l'on admette que la théorie de la science doit refléter son exercice, ce phénomène doit se traduire à terme dans l'épistémologie des mathématiques, si celle-ci veut être pertinente : c'est le

---

[30] Ce réalisme ontologique prend souvent la figure du "platonisme des mathématiciens" : "In my writings on the philosophy of mathematics, I have been concerned about the philosophical stance or preconceptions of practicing mathematicians, whether explicitly formulated or not. As I have written, this usually involves some choice or combination or alternation of "formalism" and "Platonism"." R. Hersch [31]. On lira les témoignages de deux grandes figures des mathématiques récentes [38, 11], qui traduisent l'intérêt général porté par la communauté à ces questions.

principal reproche que l'on peut adresser, du point de vue des mathématiques, aux théories philosophiques contemporaines de la vérité, que de refuser majoritairement de considérer les problèmes ontologiques au nom de positions de principe anti-métaphysiques.

Pour autant reste posée la question de droit : Pourquoi et comment l'existence et l'objectivité mathématique sont-elles possibles au-delà des critères de non-contradiction de la logique formalisée ? Pourquoi la vérité mathématique a-t-elle, au-delà de sa dimension booléenne, à voir avec notre intuition, ou encore avec les structures et les lois du monde phénoménal[31] ? Pour répondre avec Heidegger, l'essence de la vérité ne s'épuise pas dans les critères de véracité : cette essence est plutôt à rechercher dans les conditions de possibilité de la vérité –dans ce qui rend possible l'adéquation de la pensée à son objet. À l'image de la théorie kantienne du schématisme, il y va là d'un art caché dans les profondeurs de l'âme humaine. Il nous faut donc apprendre à interroger la vérité dans ses manifestations et dans son déploiement au sein de la pratique mathématique. Retourner, en fin de compte, aux fondements de la méthode phénoménologique et de la logique : penser, à la lumière des mathématiques contemporaines et de leur méthodologie, la vérité dans la dualité noético-noématique de la pensée et de ce vers quoi elle se dirige ; comprendre la logique sur un mode aristotélicien et frégéen, à la fois comme *Organon* et comme science des lois pures de la pensée.

BIBLIOGRAPHY

[1] Benoist J., 2002, Entre Acte et sens. La théorie phénoménologique de la signification. Paris: Vrin
[2] Benoist J., 2005, Les Limites de l'intentionnalité, Paris: Vrin
[3] Beaulieu L., 1989, Bourbaki, une histoire du groupe de mathématiciens français et de ses travaux, Université de Montréal
[4] Bégout B., 2002, "La Pensée en acte. Logique et activité chez Frege et Husserl", in R. Brisart (ed.), Husserl et Frege, Les ambiguïtés de l'antipsychologisme, Paris: Vrin, pp. 118-151
[5] Boniface J., 2003, Hilbert et la notion d'existence en mathématiques. Paris: Vrin
[6] Bourbaki N., 1970, Théorie des ensembles, Paris: C.C.L.S
[7] Bourbaki N., 1998, "L'architecture des mathématiques", in F. Le Lionnais (ed.), Les grands Courants de la pensée mathématique, Paris: Hermann
[8] J. Bouveresse, 2006, "La "thèse de l'inexprimabilité du contenu" a-t-elle été réfutée?", in Bouveresse J., Chapuis-Schmitz D., L'Empirisme logique à la limite. Schlick, le langage et l'expérience, Paris: CNRS éditions, pp. 137-162

---

[31] Il est tentant d'étendre la portée de ces question et de les confronter à la thèse de l'inexprimabilité du contenu. Rappelons que cette thèse, énoncée d'abord par Schlick, affirme que le langage n'exprime que la forme, jamais le contenu de l'expérience. Voir [8, 28]. L'analyse de B. Harrison selon laquelle "La différence entre structure et matériau, entre forme et contenu, est grosso modo, la différence entre ce qui peut être exprimé et ce qui ne peut pas être exprimé" [28] fait lointainement écho aux travaux heideggériens sur les possibilités du langage (qui vont, chez lui, bien au-delà des possibilités qui lui sont ouvertes dans la perspective de Schlick).

- [9] Brisart R. (ed.), 2002, Husserl et Frege, Les ambiguïtés de l'antipsychologisme, Paris: Vrin
- [10] Brisart R., 2002, "Le problème de l'abstraction en mathématiques", in R. Brisart (ed.), Husserl et Frege, Les ambiguïtés de l'antipsychologisme, Paris: Vrin
- [11] Connes A., 2000, Triangle de pensées, Paris: Odile Jacob
- [12] Corry L., 1996, Modern Algebra and the Rise of mathematical Structures, Bâle: Birkhaüser
- [13] Courtine J.-F. (ed.), 1996, Phénoménologie et logique, Paris: Presses de l'école Normale Supérieure
- [14] Dastur F., 1996, "La doctrine du jugement, la métaphysique du principe de raison et la logique", in J.-F. Courtine (ed.), Phénoménologie et logique, Paris: Presses de l'école Normale Supérieure, pp. 283-296
- [15] Dummett M., 1991, Les Origines de la philosophie analytique, trad. fr. M.A. Lescourret, Paris: Gallimard
- [16] Engel P., 2001, "Is Truth a Norm?", in P. Kotatko, P. Pagin, and G. Segal (eds.), Interpreting Davidson , CSLI Publications
- [17] English J., 2002, "Dans sa recension de la Philosophie de l'arithmétique, Frege a-t-il vraiment compris le projet de Husserl? ", in R. Brisart (ed.), Husserl et Frege, Les ambiguïtés de l'antipsychologisme, Paris: Vrin
- [18] Field H., 1986, "The Deflationary Conception of Truth", in G. MacDonald and C. Wright (eds.), Fact, Science and Morality, Oxford: Blackwell
- [19] Fisette D., 1994, Lecture frégéenne de la phénoménologie, Paris: Éditions de l'éclat
- [20] Fisette D., 2002, "Logique et philosophie chez Frege et Husserl", in R. Brisart (ed.), Husserl et Frege, Les ambiguïtés de l'antipsychologisme, Paris: Vrin pp. 50-73
- [21] Føllesdal D., 1958, Husserl und Frege, ein Beitrag zur Beleuchtung der Entstehung der phänomenologischen Philosophie, Oslo: Aschehoug
- [22] Frege G., 1884, Die Grundlagen der Arithmetik, Breslau: Koebner, Breslau
- [23] Frege G., 1971, Écrits logiques et philosophiques, trad. C. Imbert, Paris: Le Seuil
- [24] Frege G., 1971, "Recherches logiques 1, La Pensée", in G. Frege, Écrits logiques et philosophiques, trad. C. Imbert, Paris: Le Seuil
- [25] Frege G., 1971, "Sens et dénotation", in Écrits logiques et philosophiques, trad. C. Imbert, Paris: Le Seuil
- [26] Frege G. et Hilbert D., 1992, Correspondance Frege/Hilbert, tr. fr. J. Dubucs, in F. Rivenc et Ph. de Rouilhan, Logique et fondements des mathématiques. Anthologie (1850-1914), Paris: Payot
- [27] Haaparanta L. (ed.), 1994, Mind, Meaning and Mathematics. Essays on the Philosophical Views of Husserl and Frege, Dordrecht: Kluwer
- [28] Harrison B., 1973, Form and Content, Oxford: B. Blackwell
- [29] Heidegger M., 1954, Sur l'Essence de la vérité, Klostermann, trad. franç. A. de Waelhens et W. Biemel, 1968, Paris: Gallimard
- [30] Heidegger M., 1978, Metaphysische Anfangsgründe der Logik im Ausgang von Leibniz, Ga 26, Francfort: Klostermann
- [31] Hersch R., June 2008, "On Platonism", EMS Newsletter, pp. 17-18
- [32] Hilbert D., 1971, Les Fondements de la géométrie, Paris: Dunod, rééd. 1997, Jacques Gabay
- [33] Husserl E., 1950, Idées directrices pour une phénoménologie, trad. P. Ricoeur, Paris: Gallimard
- [34] Husserl E., 1969, Recherches logiques, trad. H. Elie, A. Kelkel, R. Scherer, Paris: P.U.F.
- [35] Husserl E., 1975, Articles sur la logique (1890-1913), trad. J. English, Paris: P.U.F.
- [36] Janicaud D., 1985, La puissance du rationnel, Paris: Gallimard
- [37] Mohanty J.N., 1994, "Husserl's "Logic of Truth"", in L. Haaparanta (ed.), Mind, Meaning and Mathematics. Essays on the Philosophical Views of Husserl and Frege, Dordrecht: Kluwer, pp. 141-160

[38] Mumford D., Dec. 2008, "Why I am a Platonist", EMS Newsletter, pp. 28-30
[39] Osborn A., 1949, Edmund Husserl and his Logical Investigations, 2nd ed., New-York-Londres: Garland Publ.
[40] Patras F., 2001, La Pensée mathématique contemporaine, Paris: P.U.F.
[41] Patras F., 2007, "Pourquoi les nombres sont-ils naturels?", in L. Boi, P. Kerszberg, F. Patras (eds), Rediscovering Phenomenology. Phenomenological Essays on Mathematical Beings, Physical Reality, Perception and Consciousness, Dordrecht: Springer
[42] Riemann B., 1898, Œuvres Complètes, Paris: Gauthier-Villars
[43] Shapiro S., 1997, Philosophy of Mathematics: Structure and Ontology, New York: Oxford University Press
[44] Wittgenstein L., 1959, Tractatus logico-philosophicus, Oxford: Basil Blackwell

Frédéric PATRAS
Laboratoire Dieudonné
Université de Nice et CNRS, France

Modal Truths From an Analytic-Synthetic Kantian Distinction

FRANCESCA POGGIOLESI

1 Introduction

The main aim of this paper is to answer the following question: are modal truths analytic? The very same type of question was posed by Hintikka [3] with respect to first-order logical truths (while, as far as we know, nobody has ever wonder about the analyticity of modal logic). Hintikka showed that the answer, contrary to what one might think at the first glance, is far from being trivial. In order to develop our answer, we will closely tread the same path as Hintikka.

Hintikka distinguishes three notions of analytic:

(I) analytic truths are sentences that are true by sole virtue of the meaning of the terms they contain,

(II) analytic truths do not convey any factual information,

(III) analytic truths can be shown to be true by strictly analytic methods.

Let us get a closer look to these different notions of analyticity. The first notion of analytictiy has to be completely disregarded. Indeed, the attacks raised by Quine ([11]) towards such notion have clearly shown that it is, as it stands, unsatisfactory. Moreover it makes all the logical truths trivially analytic, and it is therefore irrelevant to our purposes. So, let us turn our attention to definitions II and III of analyticity. Following Hintikka, (a certain part of) first-order logic is analytic in the sense II of analyticity, but synthetic in the sense III of analyticity. Our goal is to show that modal logic is also synthetic in the sense III of analiticity. We will not deal with the question of whether modal logic is analytic in the sense II of analyticity. Such a question can be the subject of future research.

Let us start our task by analysing the notion III of analyticity.

2 Analytic by means of analytic methods

Let us ask what can be said of the sense of analyticity defined by III. Here it is advisable to first precisely define the concept of analytical argument-step and then to extend the definition to the larger concept of argument. The basic idea of sense III seems to be expressible as follows:

> III(a) All that is said by the conclusion of an analytic argument step is already said in the premise(s)

A conclusion of an analytic argument-step merely repeats what has already been explicitly stated or simply mentioned in the premise(s). This seems to be the more standard and common sense of the word *analysis*. Therefore our definition III(a) seems to be correct. Although correct, definition III(a) is admittedly very vague. Hence it is our purpose to make it somewhat clearer. One way of making it clearer, it is to formalise it. Despite the precision and transparency that a formalisation gives us, we will not go in this direction. Indeed we aim at being clear and general at the same time; unluckily a formalisation would oblige us to choose a precise formalism, and therefore to lose the broader view on the notion of analyticity. So we have to find another solution. We propose the following one:

> III(b) In the conclusion of an analytic argument step no more objects are considered together at one and the same time then were considered together in the premise(s)

In order to clarify the meaning of analytic argument step, we use the notion of objects and their interrelations. For the conclusion to merely repeat what is said in the premise(s), the number of objects and their interrelations occurring in the conclusion must be the same as the number of objects and their interrelations occurring in the premise(s). We underline that in [3, p. 180], Hintikka proposes a very similar notion of analytic-argument step: the only difference being that wherever we use the term "object", he uses the term "individual". We think that the term "object" is more general, and thus make definition III(b) applicable not only to first-order logic, but also to modal logic. In the next section we will specify what we exactly mean by number of objects and their interrelations.

We have thus elucidated the notion of analytic-argument step. But our goal was to define the broader notion of argument. For this, it suffices to generalize what we have said up to this point. We can claim that a proof of q from p is analytic in sense III(b) if no more objects are considered at any of the intermediate stages than are already considered either in p or in q. A modal true sentence p will be said analytic if it can be proved to be true by strictly analytic means, i.e. by an argument where no more objects are considered together than in p.

Let us conclude this section by relating definition III(b) with Kant's distinction between analytic and synthetic (see [4, 5, 6]). In order to explain such a bound, let us firstly relate the sense III(b) of the term "analytic" to the sense that this word has traditionally had in geometry. A geometrical argument can be called analytic in so far as no new construction is carried out in it, i.e. in so far as no new lines, points circles and their like are introduced during the argument. On the other hand, a geometrical argument is said to be synthetic if these new entities are introduced. Let us compare this notion of analytic with the notion introduced in III(b). It seems enough clear that the former notion is just a more specific version of the latter notion. Indeed in the notion III(b) we generally say that in an analytic argument step, no new object should be introduced passing from the premise to the conclusion; in the geometrical notion, we specify what kind of object (i.e. point, lines, circles) should not be introduced passing from the premise to the conclusion.

According to Hintikka [2], Kant's usage of the term "analytic" and "synthetic" largely follows the geometrical paradigm. Therefore Kant's usage of these terms comes pretty close to the sense that these terms have in definition III(b).

3 First-Order Logic

As we have already said in the introduction, Hintikka deals with the question of whether the sentences of first-order logic are analytic in the sense III of analyticity that we have just explained. Let us dedicate this section to a presentation of Hintikka's solution.

First of all, let us introduce the key notion of *degree* of a sentence. In order to define the degree of a sentence, we have to introduce the notion of *depth* of a sentence. The depth of a sentence A, in symbols $d(A)$, can inductively be defined in the following way:

$d(A) = 0$, if A is an atomic sentence or an identity

$d(A_1 \wedge A_2) = d(A_1 \vee A_2) = \max\left((d(A_1), d(A_2))\right)$

$d(\exists x A) = d(\forall x A) = d(A) + 1$

In intuitive terms, the depth of a sentence is nothing but the maximum number of quantifiers whose scopes all overlap in it. For instance, we have: $d(P(a,b)) = 0$; $d(\exists x P(x,a)) = d(\exists x P(a,x)) = 1$; $d(\exists x P(a,x) \wedge \exists x P(x,a)) = 1$; $d(\forall x(\exists y P(x,y) \vee \exists y \exists z(P(y,z) \wedge P(z,x)))) = 3$.

Once the notion of depth of a sentence clarified, we can introduce the *degree* of a sentence. The degree of a sentence corresponds to the sum of its depth plus the number of free individual symbols occurring in it (constants or free variables).

The notion of degree of a sentence will be important in what follows, so it is worth explaining it carefully. The degree of a sentence serves to identify the number of objects (in case of first-order logic, the number of individuals) whose properties and interrelations one considers (or might consider) in that sentence. Of course this number includes the individuals referred to by the free individual symbols of the sentence. It also includes all the indefinite individuals introduced by the quantifiers within the scope of which we are moving. It does not include any other individual. The maximum number of all these individuals is just the degree of the sentence in question, which is therefore the maximum number of individuals we are considering together in the sentence.

Recall now definition III(b). In that definition, we used the expression "objects considered together at one and the same time". We promised to clarify this expression. The notion of degree serves this goal. Indeed the degree of a sentence is nothing but the number of objects considered together at one and the same time in that sentence.

As we have just said, Hintikka claims that certain sentences of first-order logic are not analytic in the sense III(b) of analyticity. Now we have all the means to explain how Hintikka justifies his claim. Let us then consider the rule of existential instantiation of the natural deduction calculus for first order logic. Such a rule has the following form:

$$\begin{array}{c} d \\ \vdots \\ \underline{\exists x Px} \\ P[a/x] \end{array} \; EI$$

where a is a free individual symbol that should not have been used elsewhere in the derivation d and $P(a/x)$ is the result of replacing x by a in P (wherever it is bound to the initial quantifier $\exists x Px$).

The rule of existential instantiation is a synthetic rule: if we take objects to be individuals, it increases the number of objects considered together at one and the same time by adding a new one. Indeed in the rule EI we pass from the formula $\exists x Px$ to the formula $P[a/x]$, hence, thanks to the rule EI, we might pass from the formula $\exists x Px$ to the formula $\exists x Px \to P[a/x]$: while the formula $\exists x Px$ has degree 1, the formula $\exists x Px \to P[a/x]$ has degree 2. Therefore the rule of existential instantiation is synthetic because it allows to infer a conclusion which has a degree higher than that of the premise(s). Therefore the first-order logic sentences that are provable by using this rule are synthetic because they cannot be proved by strictly analytic methods. An example of this type of sentences are the laws of exchanging adjacent quantifiers.

Given this situation, one could naturally ask whether the conclusion that has just been drawn is not an accidental peculiarity of natural deduction. One might think that another formalism could allow to prove all the first-order sentences in a purely analytic way. Unluckily it is not so. Let us consider the rule that eliminates the existential quantifier in tableaux calculi, i.e. the exposition rule:

$$\exists x Px$$
$$|$$
$$|$$
$$P(x/a)$$

where a is a constant that has not been used elsewhere in the derivation. Even in this case, the degree of the conclusion is higher in the degree of the premise. The same happens in the sequent calculus with the rule that introduces the existential connective on the left side of the sequent[1]

$$\frac{P(x/a), M \Rightarrow N}{\exists x Px, M \Rightarrow N}$$

All these rules are synthetic. This is a feature of every complete proof procedure in quantification theory. Indeed every such proof procedure makes use of sentences of higher degree than that of the sentences to be proved. This way Hintikka concludes that certain sentences of first order logic are synthetic in the sense III(b).

4 Modal Logic

Let us now focus on modal logic. As we have already said our aim is to show that certain modal logical truths are synthetic in the sense III(b) of analyticity. In order to draw this conclusion, let us proceed as follows. Let us first of all remind the reader of an important result that links modal logic to first-order logic.

---

[1] We take for granted that the reader knows that this rule corresponds to the exposition and the instantiation rules in tableaux calculi and natural deduction calculi, respectively (for further detail see, for example, [12]).

Definition 4.1. Let x be a first-order variable. The standard translation ST_x taking modal formulas to first-order formulas is defined as follows:

- $ST_x(p) = Px$
- $ST_x(\bot) = x \neq x$
- $ST_x(\neg A) = \neg ST_x(A)$
- $ST_x(A \vee B) = ST_x(A) \vee ST_x(B)$
- $ST_x(\diamond A) = \exists y(Rxy \wedge ST_y(A))$

where y is a fresh variable (that is, a variable that has not been used so far in the translation).

It should be clear that this definition makes good sense: it is essentially a first-order reformulation of the modal satisfaction definition. For any modal formula A, $ST_x(A)$ will contain exactly one free variable (namely x); the role of this free variable is to mark the current state; this use of a free variable makes it possible for the global notion of first-order satisfaction to mimic the local notion of modal satisfaction. Furthermore, observe that modalities are translated as bounded quantifiers, and in particular, quantifiers bounded to act only on related states; this is the obvious way of mimicking the local action of the modalities in first-order logic. Finally it is clear from this translation that there is a strict correspondence between the existential quantifier of first-order logic and the modal operator $\diamond$ (as there is a strict correspondence between the universal quantifier and the $\Box$).

Given this correspondence, and taking into account what we have previously shown for first-order logic through the rule that eliminates the existential quantifier in natural deduction, let us consider the rule that eliminates the diamond in the natural deduction calculus for modal logic. Unluckily, as far as we know, it does not exist a sort of standard natural deduction calculus for modal logic. Therefore we do not have a standard rule for the elimination of the diamond operator that we can refer to. The problem lies in the fact that proof theory for modal logic is still a pretty young enterprise and that the research has obtained better results in the sequent calculus framework. Let us then have a look to the rule that introduces the diamond on the left side of the sequent in a sequent calculus. In this case, we have different rules at our disposal. Each rule belongs to a different calculus, and each calculus for modal logic is a different generalisation of the standard sequent calculus. The latest extension of the Gentzen sequent calculus is the tree-hypersequent method (see [10]). In what follows, we will mainly deal with this formalism since it is the best suited for our purpose, but we will also make reference to the other formalisms.

Let us try to explain the tree-hypersequent method in the most general and least formal possible way (for a more accurate and formal description see [10]). First of all, in the tree-hypersequent method, instead of dealing with one sequent a time (as is the case in the classical sequent calculus), i.e. with objects of the form

$$M \Rightarrow N$$

where M and N are multisets of formulas, we deal with n sequents a time. These n sequents, that are standardly called *hypersequents*, are arranged in such a way that they can form a tree; hence they are named *tree-hypersequents* There exists a semantic way to look at tree-hypersequents: they can be seen as tree-frames of Kripke-semantics, where each sequent represents a world of the tree-frame. The rule that introduces the diamond on the left side of the sequent in the tree-hypersequent setting is the following one:

$$\frac{G[M \Rightarrow N/ \Rightarrow A]}{G[M \Rightarrow N, \diamond A]} \diamond L$$

The rule $\diamond L$ should be read (bottom-up) in the following way: in a tree-frame G, if there is a world x (denoted by the sequent $M \Rightarrow N, \diamond A$) that contains the formula $\diamond A$, then we can construct a new world y (denoted by the sequent $\Rightarrow A$) such that xRy (the relation xRy is denoted by the slash) and y contains the formula A. What can easily be noticed by looking at the rule $\diamond L$ is that in the passage from the premise to the conclusion (reading the rule bottom-up), we add some new structure, namely a slash and a new sequent. Let us further examine what this addition of structure involves.

From the standard translation of Definition 4.1., we know that there exists a strict link between the possible words of modal logic and the individuals of first order logic. Both are the objects that the first-order and modal sentences talk about. In first-order logic, in order to be able to identify these objects and their interrelations, we have introduced the notion of degree. We have to introduce a similar notion for modal logic. We do it by defining the concept of *mdegree* in the following way. The mdegree of a sentence A, in symbols $md(A)$, is the number of sequents occurring in the longest branch of a tree-hypersequent. A brief reflection will allow the reader to realise that the notion of mdegree of a modal sentence tries to capture the same idea that is at the base of the notion of degree i.e. that of counting the number of objects and their interrelations (in the modal case, related worlds) that are represented by a sentence.

Thanks to this notion of mdegree, we are now able to check whether the rule $\diamond L$ is a synthetic rule, according to definition III(b). Indeed it is: the rule $\diamond L$ allows us to infer, reading it bottom-up, a conclusion with a mdegree higher than the premise. We can then draw the conclusion that the modal truths that are provable by means of this rule are synthetic in the sense III(b) of the distinction between analytic and synthetic sentences.

Let us then briefly sum up what has been said up to now. Between modal logic and first order logic there is an evident parallelism that can even be formalised by a rigorous definition. Sentences of modal logic talk about worlds and their interrelations, as sentences of first-order logic talks about individuals and their interrelations. In the case of first-order logic, the number of objects and their interrelations expressed by a sentence is calculated by summing up the free individuals occurring in the sentence plus the number of quantifiers whose scope all overlay in the sentence. In the case of modal logic, thanks to the tree-hypersequent method, the number of objects and their interrelations expressed by a sentence is much more easily calculated by counting the number of sequents occurring in a the longest branch of a tree-hypersequent. In the rule of existential instantiation, as well as in the rule $\diamond L$, the number of objects considered together at one and the same time increases from the premise to the conclusion. This fact allows us to claim that both rules are synthetic, and that the sentences of first-order logic and modal logic, respectively proved by means of the rule of instantiation and the rule $\diamond L$, are synthetic too according to definition III(b).

Let us conclude this section by briefly seeing other formalisms where we have a rule that introduces the diamond on the left side of the sequent. For instance, let us consider the display method, as well as the multiple sequent method (for an accurate description of these two generalisations of the sequent calculus see [10] and [13]). In both these cases we have the following rules:

$$\frac{A \Rightarrow \bullet N}{\diamond A \Rightarrow N} \diamond L* \qquad \frac{A, M \diamond \Rightarrow N}{\diamond A \Rightarrow N} \diamond L**$$

where the symbol $\bullet$ is the characteristic structural connective of the display method, while $\diamond \Rightarrow$ is the characteristic structural connective of the multiple sequent method. Without dwelling on the interpretations of the two new structural connectives of the display method and the multiple sequent method, which would lead us too far away from our principal goal, it is easy to notice that in the two rules $\diamond L*$ and $\diamond L**$ we can observe the same phenomenon that we have underlined in the rule $\diamond L$, i.e. a new structural element appearing. This new structural element could be taken as the new introduced object that renders the rules concerning the diamond all synthetic. Hence these examples clearly show that the fact that certain modal sentences are synthetic do not depend on a particular formalism,[2] but it is, on the contrary, an unavoidable feature of modal logic, as it was the case for first-order logic.

5 First-Order Logic and Modal Logic

We have thus shown that certain modal sentences are synthetic because provable by synthetic means. Note that such a conclusion could also serve to shed new lights on the relations between first-order logic and modal logic. Let us first of all remind the reader the following theorem, which is based on Definition 4.1

Theorem 5.1. Let A be a modal formula, then:

for all models $\mathfrak{M}$, $\mathfrak{M} \models A$ if, and only if, $\mathfrak{M} \models \forall x ST_x(A)$

In other more informal words, Theorem 5.1 states that modal formulas are equivalent to first order formulas in one free variable. On purely syntactic grounds it is obvious that the standard translation is not surjective (standard translations of modal formulas contain only bounded quantifiers). The question is then: could every first-order formula (in the appropriate correspondence language) be equivalent to the translation of a modal formula? No. This is very easy to see: whereas modal formulas are invariant under bisimulations, first-order formulas need not be; thus any first-order formula which is not invariant under bisimulations cannot be equivalent to the translation of a modal formula (for further details see [1]). This means that modal logic is a subset of first-order logic, under an adequate translation. Given this situation, it could have been that the first-order formulas that we have classified as synthetic were exactly the first-order formulas not translatable in modal logic. In that case modal logic would have been the analytic subset of first-order logic. In this paper we have on the contrary shown that certain modal formulas are synthetic and therefore that the modal subset of first-order logic contains (at least) some of the synthetic first-order formulas.

---

[2]Note that there also exist standard sequent calculi for modal logic, e.g. [8] and [9]. In these calculi, the rule that introduces the diamond on the left side of the sequent is analytic, i.e. it does not introduce any new structure. On the other hand, in these calculi the cut-rule, the non analytic rule *par excellence*, is not eliminable, therefore the lack of analyticity is there anyway.

6   Kant and the analytic-synthetic distinction

In Section 2 we have explained the distinction between analytic and synthetic from a geometrical, and more generally, from a mathematical point of view. We have also briefly said that Kant was very close to this kind of distinction and therefore to the sense of analyticity explicitly stated in III(b). In this Section we will further examine these issues in the light of the conclusions that we have obtained in Section 4.

Let us go back to Euclid and Aristotle. A reason for uniting these two illustrious thinkers lies in the fact that they both used the word, *echtesis*, for denoting two procedures that seem, at the first sight, pretty different. On the one hand, we can call echtesis, or exposition, the part of an Euclid's theorem where a new figure is introduced or drawn for the first time. On the other hand, we can call echtesis, or exposition, a procedure, used in syllogistic theory, that pretty much resembles to the rule of existential instantiation (for further and more precise references see [7]). So we have (i) a geometrical exposition, (ii) a syllogistic exposition, and (iii) an inferential exposition represented by the rule that eliminates the existential quantifier in first-order logic. The crucial common point of these steps is that they all introduce a new element. As we have seen in the last section, even in modal logic we have a similar situation. When it come to the diamond operator, we introduce in the derivation some structure that was not present in the premise(s). What Kant would have thought of these four different procedures? As it has been suggested in [2], Kant would have probably indicated them as synthetic procedures. Indeed Kant seems to claim that mathematical truths are synthetic because they are based on the use of constructions. But construction is exactly what characterise (i)-(iii) plus the modal rule $\diamond L$. In the rule $\diamond L$, when we pass from the conclusion to the premise, we literally construct a new world-sequent linked to the others by a relation-slash. In (ii) and (iii) the construction is in terms of introduction or "exhibition" of an individual idea to represent a general concept. In (i) the construction is done by drawing a new figure. So for Kant these fours procedures would have been perfect examples of synthetic modes of reasoning in mathematics. In particular (iii) and $\diamond L$ would have led him to agree with us that certain first-order logic sentences, as well as certain modal sentences are synthetic because provable by synthetic means.

BIBLIOGRAPHY

[1] P. Blackburn, M. de Rijke, and Y. Venema. *Modal Logic*. Cambridge University Press, Cambridge., 2001.
[2] J. Hintikka. Kant's theory of mathematics. *Ajatus*, XXII:5–85., 1959.
[3] J. Hintikka. Are logical truths analytic? *The Philosophical Review*, 74:178–203., 1965.
[4] E. Kant. *Critique of Pure Reason*. Cambridge University Press, Cambridge, 1999.

[5] E. Kant. *Prolegomena to Any Future Metaphysics*. Hackett, Indianapolis, 2001.
[6] E. Kant. *Critique of Practical Reason*. Cambridge University Press, Cambridge, 2010.
[7] J. Lukasiewicz. *Aristotle's Syllogistic*. Oxford University Press, Oxford, 1951.
[8] M. Ohnishi and K. Matsumoto. Gentzen method in modal calculi. *Osaka Mathematical Journal*, 9:113–130, 1957.
[9] M. Ohnishi and K. Matsumoto. Gentzen method in modal calculi II. *Osaka Mathematical Journal*, 11:115–120, 1959.
[10] F. Poggiolesi. *Gentzen Calculi for Modal Propositional Logic*. Trends in Logic Series, Springer, 2010.
[11] W. V. O. Quine. *From a Logical Point of View*. harvard University Press, Harvard, 1953.
[12] A. S. Troelestra and H. Schwichtenberg. *Basic Proof Theory*. Cambridge University Press, Cambridge, 1996.
[13] H. Wansing. *Displaying Modal Logic*. Kluwer Academic Publisher, Dordrecht, 1998.

Francesca POGGIOLESI

UMR 8590 IHPST - Institut d'Histoire et de Philosophie des Sciences et des Techniques

Université Paris 1 Panthéon-Sorbonne, CNRS, ENS, France

How to Hintikkize a Frege

FABIEN SCHANG

1 Frege's theory of meaning and its critics

1.1 A one-sorted semantics

Frege's rationale is to uphold his own theory of meaning by means of an argument by contraposition $(\alpha \supset \psi) \supset (\neg \psi \supset \neg \alpha)$. It runs as follows: If

(α) every sign (*Zeichen*) of an arbitrary sentence includes both a sense (*Sinn*) and a reference (*Bedeutung*),

then

(ψ) no substitution of equivalent components (with the same reference) alters the whole.

Now if

($\neg\psi$) at least one counterexample of this substitution *salva veritate* can be found,

then

($\neg\alpha$) Frege's theory of meaning collapses.

The kinds of argument under review are the following: Frege asserts both α and ψ; Kripke denies α and asserts ψ; Hintikka asserts α and denies ψ.

> **FIRST OUTLINE OF FREGE'S THEORY OF MEANING**
>
> One-sorted semantics + Compositionality
> =
> extensional (standard) logic

The contrast between two modes of meaning (conveying information) through every linguistic item, that is the one between "sense" and "reference", primarily cancels some naïve theory of meaning, according to which every sign would purport to tag or label one object (recall Quine's museum myth of meanings [13], or Saint Augustine's assimilation between meaning and pointing something out [14]). How to account for the information gain from "$a = a$" to "$a = b$", if meaning is reduced to reference?

For this purpose, Frege's philosophy of language relies upon two main assumptions: compositionality, and extensionalism. According to the former, the reference of a complex sentence is determined by the reference of its components and sense (reference) cannot contribute to determinate a reference (sense). A crucial point for the following: the references of sentences are truth-values, among the True and the False. It is for this reason that Frege's logic, because it has primarily occurred as a science of truth, is uniquely concerned with references and never deals with matters of sense concerning the informative value conveyed by a sign. The close connection between logic and truth has been famously claimed by Frege [4: 170] in these introductory words:

> The word "true" indicates the aim of logic as does "beautiful" that of aesthetics or "good" that of ethics. All sciences have truth as their goal; but logic is also concerned with it in a quite different way from this. It has much the same relation to truth as physics has to weight or heat. To discover truths is the task of all sciences; it falls to logic to discern the laws of truth.

Because logic deals with truth, and truth is a reference, the very idea of a logic of sense cannot be but troublesome. Accordingly, a way out is to revisit the logical notions of truth and sense.

Following his one-sorted semantics, every sign expresses a sense and refers to an object: individual variables ($x, y, z, \ldots$) refer to individuals while expressing an individual concept; predicate variables ($P, Q, R, \ldots$) refer to properties while expressing concepts; sentence variables ($p, q, r, \ldots$) refer to truth-values while expressing "propositions". Every sentence (*Satz*) includes both a thought as its sense (the *Sinn*, as the *Gegebenheitsweise* or way of giving references) and a truth-value as its reference (the True or the False, among its two possible objects). For according to Frege, the thought (*Gedanke*) is the sense of a sentence and corresponds to the other name of the proposition. In this sense, thought is an objective item of language and does not constitute a private representation (a "*Vorstellung*"). Now the process of internalization leads to an essential problem for analytical philosophy, concerning the boundary between objective and subjective domains. To put it in words: how far can we turn subjective contents of thought into objective terms of logic? We return to this point in the end of the paper.

Meanwhile, Frege's theory of language entails that any two sentences with the same truth-value refer to the same thing; these merely differ by their sense. According to the so-called "rule of substitution" that prevails for every sign designating identical objects, it results in three variants of extensionalism — this is Frege's second main assumption — for three distinct sorts of terms (individuals, predicates, and sentences):

Leibniz's Law: $\forall x \forall y \forall P[(x = y) \supset (Px \equiv Py)]$
Principle of Coextensivity: $\forall P \forall Q \forall x[(P = Q) \supset (Px \equiv Qx)]$
Principle of Extensionality: $\forall p \forall q \forall \delta[(p \equiv q) \supset (\delta p \supset \delta q)]$ (where δ is a context variable)

Is this formal theory of language reliable in any case? As is known, there seems to be non-extensional, i.e. intensional contexts δ in which the answer is negative. Let us consider the case of referential opacity in belief contexts. Although Tom believes that Cicero wrote *De Senectute* (symbols: $B_{Tom}F(a)$), and Cicero denoted the same invididual as Tullius (symbols: $a = b$), Tom may not believe that Tullius wrote *De Senectute* (i.e. $\neg B_{Tom}F(b)$). A counterexample seems to be found here to substitutivity of identicals, and so does accordingly for extensionalism. Formally speaking, the reasoning

$$\frac{B_{Tom}F(a) \quad a = b}{B_{Tom}F(b)}$$

does hold in Frege's extensional logics but should not in an informal reasoning of natural language.

By analogy with quotation or indirect discourse contexts, Frege's way out consists in making a distinction between direct (*gerade*) and indirect (*ungerade*) references: in a belief context, terms do not have their usual reference; sentences do not refer to truth-values any more but, rather, they express thoughts, so that the thought is that which is *expressed* in a direct context and *referred* to in an indirect context. Now following the principle of compositionality, no thought can contribute to the truth-value of a sentence, and the intensional counterexample is thus reduced to a "semiotic illusion" (i.e. a confusion between two tasks in a sentence) [3:116].

> This arises from an imperfection of language, from which even the symbolic language of mathematical analysis is not altogether free.

The contemporary intensional or non-classical logics have been largely motivated by this problem of referential opacity; now the previous Fregean argument seems to challenge their very legitimacy: don't their logical forms result from a sort of category mistake, i.e. a symbolic confusion between the sense of a sign and its reference? At least three basic features of a "semantic shift" can be actually found in the current practice of these intensional logics: a treatment of incomplete thoughts as complete thoughts; a rejection of the one-sorted semantics; an internalization of sense, notably in epistemic logic where the truth-conditions of beliefs are specified.

Would such a process have something meaningless in it?

1.2 Objections to the one-sorted semantics

Complete and incomplete thoughts

Non-classical logic is generally seen as a logic in which one of the cornerstones of Frege's theory of meaning is discarded, whether for one-sorted semantics or compositionality. Throughout [3], the author means by incomplete thoughts (as parts of thoughts) the various sorts of sentences which currently characterize such modal logics as temporal, relevant, causal or counterfactual logics.

The contrast between Fregean and modal logics relies upon the logical form to be assigned to modalities: in the case of a modal sentence δp, with an operator δ and a that-clause p, p is incomplete for Frege and complete for a modal logician. In other words, the modal logician determines the truth-value of the modal sentence according to the truth-value of its component p; Frege does not proceed in this way, because p does not have its usual reference in a modal context and, therefore, does not have any truth-value as its reference. How to account for such a difference in the logical analysis?

Let us turn again to the case of doxastic modal logic, i.e. the part of epistemic modal logic devoted to the concept of belief ("doxa"). In the sentence "Tom believes that Cicero wrote *De Senectute*, the reference of the that-clause "Cicero wrote *De Senectute*" is the sense of the whole sentence and, thus, does not determine the extension of the corresponding belief: our knowing whether Cicero did write *De Senectute* or not does not decide whether Tom does believe it or not. Syntactically speaking, the Fregean distinction between the usual and unusual reference of a sentence turns the initial sentence into a name of that-clause: the logical form of these indirect contexts of discourse is not $\delta F(a)$ but $\Delta "F(a)"$, where Δ is no longer an operator δ applied to a sentence $F(a)$ but, rather, a predicated attached to an individual name "$F(a)$"; such a transformation helps to extensionalize intensional contexts and has been used by other authors such as Carnap or Quine. In [1], for instance, Carnap handles the notion of necessity as a predicate for analyticity to which a name of sentence is attached. But the trouble with this extensional treatment concerns the quantified open sentences: the aforementioned paraphrase cannot be applied anymore, given that the transformation of an individual variable x into an individual name "x" leads to a nonsensical quantified sentence: $(\exists x)\Delta(\text{``F}(x)\text{''})$, where the relation between the bound variable x and the variable within quotes "F(x)" is like the non-semantic relation between the word "cattle" and its componing phonem "cat".

The limitation of Frege's formal theory of language needs some change, consequently: either compositionality need be abandoned, or the trouble comes from the one-sorted semantics. The latter is questioned by Kripke's theory of rigid designation. Can the problem of referential opacity be settled within such a non-Fregean semantics?

A two-sorted semantics

To the problem of substitutivity in opaque contexts, Saul Kripke [11] and such other "New Theoricians of Reference" (thereafter: NTR) as Ruth Barcan-Marcus, Keith Donnellan, Nathan Salmon, Dagfinn Føllesdal replied by a two-sorted semantics: proper names have a reference but no sense (a reminiscence of John Stuart Mill's theory of proper names), so that not every term includes both a reference and a sense. In order to emphasize this univocal relation between a proper name and its reference, Kripke and Barcan Marcus endorsed the thesis of necessity (N) for identity sentences:

$$\forall x \forall y[(x = y) \supset N(x = y)]$$

Let us borrow an example from alethic modal logic, where the central modalities are those of necessity, possibility, or contingency. If 9 is necessarily greater than 7, and 9 is the number of planets, does it entail that the number of planets is necessarily greater than 7? Unless a contingent truth of astronomy turns out to be a necessary truth, we are led to conclude that the following inference does not hold and falsifies extensionalism:

$$\frac{N(9 > 7) \quad 9 = \text{the number of planets}}{N \text{ (the number of planets} > 7)}$$

However, the theory of rigid designation sustains the substitutivity of identicals when designated by a *proper name*: it claims that a proper name has the same reference whatever the context of discourse may be (i.e. in every possible world) and keeps contributing to the reference of the componing sentence. The failure of the above inference is due to the fact that one of the substituted terms is not a proper name but, rather, a *definite description* (viz. "the number of planets"); the success condition of substitution depends upon the choice of the designating sign, and not only the designated object.

Assuming that this two-sorted semantics does work for necessity contexts, how to avoid the failure of substitutivity in intensional contexts like belief? In such a case, the rigidity of proper names and the view of a unique referring task for proper names don't seem to be sufficient: the substitution of proper names needn't preserve the truth-value (the reference) of the modified sentence. Note also that Hintikka was a conspicuous opponent of NTR, claiming that the reference of proper names is not "tagged on their forehead" but results from the very process of individuation performed by the epistemic agent [9]. At the same time, one can agree with Smith [15] that such a critic may seem to be unfair, given that Kripke's theory merely concerned the peculiar context of *alethic* modalities.

After Frege and Kripke, Hintikka's position appears to be more intuitive when applied to doxastic contexts: it maintains Frege's one-sorted semantics while rejecting its extensional principle of substitutivity. But then how to account for such a seemingly inconsistent stance, both asserting α and denying ψ in the initial Fregean implication $\alpha \supset \psi$? The following shows that the implicational form (i.e. the scheme $\alpha \supset \psi$) in the reasoning pattern of [3] relies upon a Fregean theory of truth that is not shared by Hintikka.

2 Hintikka's internalization of Frege's sense

By contrast to the supporters of NTR, Hintikka does not argue for a two-sorted semantics and sticks to Frege's one-sorted version. Yet his view of logic gives rise to two major alterations, namely: (1) the use of a "possible-worlds semantics", as a co-univocal relation (one-many) between language and the world and according to which Frege's *Sinn* comes to be synonymous with a multiple reference [6]; (2) the claim of an affinity between Frege's judgement-stroke, assertion, and belief [5]. The status of Frege's judgement-stroke remains an open question, however. For example, while taking its psychological, performative or illocutionary reading into account, Greimann [5:215] notes that

> According to Wittgenstein, the sign ⊢ is logically quite meaningless, because "in the work of Frege (and Russell) it simply indicates that these authors hold the propositions marked with this sign to be true" (cf. Wittgenstein (1921, sentence 4.442)). On this interpretation, the judgement-stroke is a psychological operator whose linguistic function is to express certain propositional attitudes.

Rather, Greimann opposes an alternative reading of the judgement-stroke as a truth-operator marking the value of the judgeable content; the present paper follows [10] and maintains the psychological version which associates the propositional attitude of belief to the judgement-stroke.

2.1 Frege's sense as referring

Valuation in a possible-world (*à la* Hintikka) semantics consists in assigning a reference (a truth-value) to a sentence among a number of possible worlds, those worlds standing for sets of sentences compatible with an agent's beliefs. In other words, any agent believes that p if and only if p is logically compatible with each of these (com)possible worlds; whereas she does not believe that p whenever, in at least one of these worlds, p is not compatible with the other sentences (i.e. whenever p denotes the False).

Such a process turns Frege's theory of truth upside down: the Fregean connection between reference and sense comes to be translated into a connection between single and multiple reference, and Hintikka's epistemic logic affords a formal device to account for the notion of sense in terms of multiple reference or truth-values assignment. To assign a reference in a world (a model) is to give a sentence its sense. The outcome of this multiple valuation is a kind of individuation for linguistic items, and the function purported to accomplish such a task does apply to the Fregean *Sinn*. For let f be an individuating function applied to references in models; then the way of givenness (*Gegebenheitsweise*) characterizing sense results in the mapping $f(w_k, p) \mapsto \{0,1\}$, where k is the label distinguishing the different possible worlds w and $\{0,1\}$ the bivalent pair of references for sentences (i.e. their truth-values).

Despite this technical formalization in terms of possible worlds, Hintikka *does* still agree with Frege's views about the two preceding theses. That is: one-sorted semantics and compositionality are still in order in his formal approach of beliefs, and the unusual reference of a term (its sense) cannot determine its usual one. A clear difference occurs in the terms employed to define the unusual reference, i.e. the sense of a sentence within a doxastic context of discourse: Hintikka associates truth-values to that which constitutes the sense of this sentence, while Frege would have taken this manœuvre to be absurd since the True and the False cannot be senses from his nonmodal perspective. Where does the difference lie between Hintikka's and Frege's theories of truth? Not in their theory of truth properly speaking but, rather, in their theories of naming truth, i.e. in the category of terms which can be said to be true. According to Hintikka, the difference is to be expressed formally by rejecting the one-world assumption [8]: there is not only one but several domains of reference to be taken into account for a semantic assessment of thoughts (i.e. beliefs, in Hintikka's epistemic parlance).

2.2 Belief as holding true

The vernacular link between belief, assertion, and holding true does justice to what Hintikka realized with his formal system, namely: a formalization of sense, depicting the process of (making) sense – giving a reference – within a logical calculus. Indeed, a close connection can be seen between the notion of belief as truth-value assignment in possible worlds and Frege's notion of assertion as the act of holding true. Do we have the same reading of "truth" in both cases, and what of truth in a possible world as a variable reference among several ones (other accessible worlds)? If everything is all right far, then sense can be determined by truth-values in epistemic logic; however, these do not stand for one and the same object referred by a sentence, as was the case with Frege's semantics (the Fregean truth-value occurs in a unique available world, i.e. the real world). Now according to what Frege meant by sense as a public, non-psychological item, the individuating function or intension advocated by Hintikka to pick out individuals within possible worlds can be rendered in Fregean terms of sense *and references* (in worlds). In this respect, Hintikka agrees with Frege but is at odds with him as to the manyworld assumption; furthermore, belief results from an individuating function based on the very notion of *Sinn*, so that any belief embedded in a judgement *is* the "thought" denoted by sentences (isn't belief a form of thought, after all?).

On the other hand, Hintikka accounts for referential opacity in terms of variable modes of reference: substitutivity fails whenever a given term fails to be given one and the same reference (truth-value) in each possible world compatible with the speaker's belief, i.e. when it is not individuated.

Let us note finally that, following van Heijenoort [16], the discrepancy between the supporters of logic as a universal language (the early Wittgenstein, Quine, Russell, Frege) and logic as a calculus (Skolem, Löwenheim, Peirce, Hintikka, Kripke) helps to mark the contemporary history of logic as a deep disagreement between universalists and algebrists. However, such a partition in the history of modern logic has been questioned by Peckhaus [12]: according to the latter, van Heijenoort would have wrongly claimed that the champions of logic as a calculus ignored the use of *quantifiers* while these already occurred in the algebrist works of Schröder. At the same time, van Heijenoort's historiographic dichotomy still makes sense according to us from another perspective, viz. the model-theoretic perspective of many-worlds assumption compatible with the algebrist's view, whereas the universalists would have been unable to think about several models because of their one-world assumption. If such a distinction between one- and many-worlds assumption is taken into account to explain the impossibility of formalizing Frege's sense, the result is the following revised version of his theory of meaning:

SECOND OUTLINE OF FREGE'S THEORY OF MEANING

One-sorted semantics + Compositionality + One-world assumption
=
extensional (standard) logic

If belief is a function determined by a set of reference assignments, and each of these assignments results in a sense, then reference can be taken to be a basic notion of semantics. Therefore, the process of internalization has helped to go upstream and account for the rise of belief in functional terms; can it now go on computing the very givenness of sense, or does such an operation remain beyond the realm of analysis? This should go beyond the area of logicians, and a division of labour occurs here between epistemic logic and epistemology. Can the following problem, i.e. how to acquire a method of individuation for the objects of discourse, be treated by a logical calculus?

3 Conclusion: the limits of internalization

The answer to the question "Can a logic of sense make sense?" is a qualified "Yes, but": a logic of Fregean sense does make sense, provided that the unique domain of reference of Frege's logical theory be modified. Contemporary formal semantics clearly corroborated this increasing use of formalism within theories of language, as witnessed by the actual prominence of the so-called non-classical or intensional logics. How far can such an increase be accepted? Can we hope a purely formal theory of the rules of meaning for ordinary language?

Michael Dummett blamed the philosophy of language for not motivating the operations at work in formal semantics [2:92]:

> A semantic explanation, entirely formulated with the help of the concept of [reference], adequately shows how a truth-value is determined by its componing words and the way in which these are composed. Yet the semantic explanation is lacking, for it does not go backwards enough: it posits a connection between its primitive symbol and an appropriate denotative, without telling us how such a connection is established. In logic, this is not mandatory; for a theory of meaning, it is essential.

Does epistemic logic bring this drawback that Dummett takes to be essential? Not as it was devised by the realist-minded logician Hintikka: it merely provides for a mathematical structure intended to describe the prefabricated function of sense, but without telling anything about its very constitution. Yet it would be worthwhile to have a closer look at the teachings of an antirealist semantics, in order to account for the modes of meaning before referring and attempt a formalization of the process of assertibility; that is: the conditions under which a subject is in position to assign a reference to a term should be taken into account. These details do not occur within a realist semantics, since the truth-conditions of a sentence transcend the cognitive skills of subjects. Among these skills is the capacity to recognize one and the same reference through a number of distinct expressions used by the same individual (i.e. the capacity to perceive the reference *transparently*).

The debate around internalization is nothing less than a debate about the place of logic within the philosophy of language. A debate that remains open to the moving history of the discipline.

Bibliographie

[1] Carnap, R. *Logische Syntax der Sprache*, Wien: Julius Springer, 1934.
[2] Dummett, M. Frege: *Philosophy of Language*, Harvard University Press, 1993.
[3] Frege, G. "Sens et dénotation", in *Ecrits logiques et philosophiques*, Paris, Seuil, collection "L'ordre philosophique", 1971, translated by Claude Imbert (original text: "Über Sinn und Bedeutung", in *Zeitschrift für Philosophie und philosophische Kritik* **100**, 1892, 25-50).
[4] Frege, G.: "La pensée", in *Ecrits logiques et philosophiques*, Paris, Seuil, collection L'ordre Philosophique, 1971, translated by Claude Imbert (original text: "Der Gedanke. Eine logische Untersuchung", in *Beiträge zur Philosophie des deutschen Idealismus* **2**, 1918-1919, 58-77.
[5] Greimann, D. "Judgement-operator as truth-operator : a new interpretation of the logical form of sentences in Frege's scientific language, *Erkenntnis* **52**, 2000, 213-38.
[6] Hintikka, J. "Modality as referential multiplicity", *Ajatus* **20**, 1957, 49-64.
[7] Hintikka, J. *Knowledge and Belief (An Introduction into the Logic of the Two Notions)*, Ithaca Press, N-Y, 1962.
[8] Hintikka, J.: "Problems of philosophy. Problem #3: one-world assumption and Frege's sense-reference distinction", *Synthese* **112**, 1997, 431-2.
[9] Hintikka, J. and Sandu, G.: "The Fallacies on the New Theory of Reference", *Synthese* **104**, 1995, 245-83.
[10] Hocutt, M. "Is epistemic logic possible?", *Notre Dame Journal of Formal Logic* **13**, 1972, 433-54.

[11] Kripke, S. *Naming and Necessity*, Harvard University Press, 1980 (1st edition: 1972).

[12] Peckhaus, V. "Calculus ratiocinator versus characteristica universalis? The two traditions in logic, revisited", *History and Philosophy of Logic* **25**, 2004, 3-14.

[13] W.V.O. Quine. *Ontological Relativity (and Other Essays)*, Columbia University Press, New York, London, 1969.

[14] Saint Augustine. *De Magistro*, in *Against the Academicians and the Teacher* (trans. Peter King), Hackett Publishing Company, 1995.

[15] Smith, D. W. "Kantifying", *Synthese* **54**, 1983, 261-73.

[16] van Heijenoort, J. "Logic as Calculus and Logic as Language", *Synthese* **17**, 1967, 324-30.

Fabien SCHANG
National Research University Higher School of Economics

The Square of Opposition that Was Lost
ANDREW SCHUMANN

1 Introduction

The square of opposition is one of the main concepts of logic. It fixes fundamental relations in logic: 1) The duality relation between conjunction and disjunction, the law of contrary, the law of *tertium non datur*. 2) The duality relation between universal quantifier and existential quantifier, the law of contrary, the law of *tertium non datur* for them. 3) The duality relation between the modal operators of necessity and possibility, the law of contrary, the law of *tertium non datur* for them and so on.

Each verb is ether informative (i.e. it describes visible action such walking, jumping) and performative (i.e. it describes action that can be never seen such as to love, hate; I do not mean gestures, because they are not actions, they are non-verbal signs to express my feeling). Informative verbs concern just singular action therefore there is no ordering relation between them. On the other hand, performative verbs express feeling and can express them weaker and harder. Accordingly, we can define the order on them. Let us notice that we ever find out a pair of performative verbs $\mathbf{F}_1$ and $\mathbf{F}_2$ with the duality relation and $\mathbf{F}_1 \supset \mathbf{F}_2$. This means that appropriate pairs of performative verbs $\mathbf{F}_1$ and $\mathbf{F}_2$ satisfy the square of opposition. Indeed, $\neg \mathbf{F}_1$ and $\mathbf{F}_2$ are in semantic opposition. This means that they both cannot be successful at the same time, however both may be unsuccessful. For example, the performative verbs 'order' and 'forbid' have such opposition. The *square of opposition* or *logical square for performative verbs* is a natural way of classifying performative verbs which are relevant to a given opposition. Starting from two performative verbs $\mathbf{F}_1$ and $\mathbf{F}_2$, the square of opposition entails the existence of two other verbs, namely $\neg \mathbf{F}_1$ and $\neg \mathbf{F}_2$. As a result, we obtain the following four distinct kinds of opposition between pairs of performative verbs, please see Figure 1:

As we see, the square of opposition is a key notion of different logics: first-order logic, modal logic, illocutionary logic. However, we claim in this paper that there are two squares of opposition. One of both is absolutely unknown still. It is said to be the "lost" square of opposition. In the paper

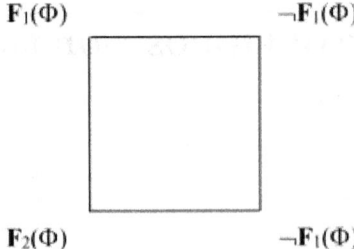

Figure 1. The conventional square of opposition for performative verbs $\mathbf{F}_1$, $\mathbf{F}_2$. For instance, $\mathbf{F}_1$ is 'bless' and $\neg \mathbf{F}_2$ is 'damn.' There can be an infinite number of such squares for different performative verbs. At least, there exists one square of opposition for each performative verb.

we prove its existence.

2 Historical background

The conventional square of opposition has been the best known logical pattern since Aristotle. However, since Kant and Rickert many philosophers have paid attention that this square does not satisfy synthetic propositions. In this paper we are proving that there are two squares of opposition under following assumptions:

- we have the Boolean complement;
- synthetic propositions cannot be reduced to manipulations with Venn diagrams, because they do not suppose including relations.

Aristotle proposed the following four oppositions: contradiction, contrariety, relation, and privation (*Categories*, Chapter 10, *Metaphysics*, Book I) that semantically underlay the square of opposition:

> We must next explain the various senses in which the term 'opposite' is used. Things are said to be opposed in four senses: (i) as correlatives to one another, (ii) as contraries to one another, (iii) as privatives to positives, (iv) as affirmatives to negatives (*Categories*, 10).

Aristotle himself described relations of the square of opposition to represent singular expressions (*Prior Analytics*, Chapter 46), see Figure 2. He had an intuition that quantifiers (both universal and existential ones) satisfy the semantic relations of the square, too:

An affirmation is opposed to a denial in the sense which I denote by the term 'contradictory', when, while the subject remains the same, the affirmation is of universal character and the denial is not. The affirmation 'every man is white' is the contradictory of the denial 'not every man is white', or again, the proposition 'no man is white' is the contradictory of the proposition 'some men are white'. But propositions are opposed as contraries when both the affirmation and the denial are universal, as in the sentences 'every man is white', 'no man is white', 'every man is just', 'no man is just (*On Interpretation*, 7).

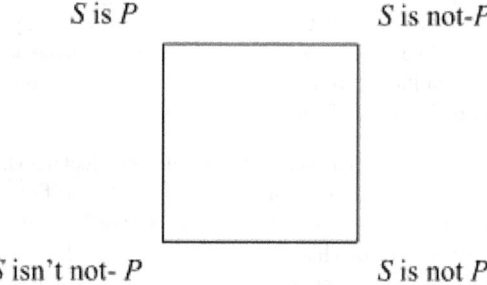

Figure 2. Aristotle's square of opposition

However, for the first time, Apuleius explicitly claimed that quantified propositions satisfy the square. He wrote a short book, the *Peri Hermeneias* [2], about logic that was used for centuries in teaching. This book is the most famous in the history of logic for including the first appearance of the square of opposition, the best known logical schema for the pedagogic purpose [2]. He considered the four oppositions: contrary, subcontrary, contradictory, subalternation. First, he described that the two *incongruae* (contrary) propositions, on the left and right sides of the top of the square, never can be true at the same time and nonetheless are sometimes false at the same time. For example, when some pleasures are good, both universal propositions are false at the same time, since it is impossible that every pleasure is both a good and not a good. The two propositions along the bottom line (i.e. the mirror-image of the contrary) are called *subpares* (subcontrary). They are never false at the same time, but they can be true at the same time. Hence, to confirm that some pleasure is a good we cannot use an argument that some other pleasure is not a good. Further, we pair together the *alterutrae* (contradictory) propositions, if we add a negation to each of the pair of alternates, e.g. not every pleasure is a good means that some pleasure is not a good. Finally, the *subalternation* appears between universal and particular propositions, when the universal implies the particular, e.g. if every pleasure is a good, then some pleasure is a good.

The meaning of propositions that satisfy the square of opposition can be checked on Venn diagrams. Recall that a Venn diagram is an ellipse that designates an extent of a concept A, i.e. a class of all real things that are denoted by A. These things are called denotations. By assumption, all inner points of ellipse designate appropriate real things. For instance, 'every man is mortal' is a true proposition, because the Venn diagram of 'man' is included into the Venn diagram of 'the mortal being' (i.e. all denotations of 'man' occur among denotations of 'the mortal being').

Kant first paid attention that there exists a true universal proposition like 'all bodies are heavy' such that Venn diagrams of its subject and predicate do not assume the including relation. So, the Venn diagram of 'body' just intersects the Venn diagram of 'heavy':

> In all judgments in which the relation of a subject to the predicate is thought ..., this relation is possible in two different ways. Either the predicate to the subject A, as something which is (covertly) contained in this concept A; or outside the concept A, although it does indeed stand in connection with it. In the one case I entitle the judgment analytic, in the other synthetic. Analytic judgments (affirmative) are therefore those in which the connection of the predicate with the subject is thought through identity; those in which this connection is thought without identity should be entitled synthetic (...) If I say, for instance, '*All bodies are extended*', this is an analytic judgment. For I do not require to go beyond the concept which I connect with '*body*' in order to find extension as bound up with it ... The judgment is therefore analytic. But when I say, '*All bodies are heavy*', the predicate is something quite different from anything that I think in the mere concept of body in general; and the addition of such a predicate therefore yields a synthetic judgment.
>
> Judgments of experience, as such, are one and all synthetic. For it would be absurd to found an analytic judgment on experience. Since, in framing the judgment, I must not go outside my concept, there is no need to appeal to the testimony of experience in its support [4].

Thus, in Kant's opinion, only analytic judgments satisfy the square of opposition without problems. For synthetic judgments Venn diagrams lose any sense and, as a result, we cannot apply the square for them. We need, as Kant affirms, the transcendental unity of apperception to subordinate them to the square. In Aristotelian logic there is the inverse relation between content (class of all connotations) and extent (class of all denotations) of a concept. By continuing the Kant's ideas, Heinrich Rickert claimed that *for synthetic judgments (propositions) there is the direct relation between content and extent of a concept.* Therefore we cannot use Venn diagrams there at all.

Thus, according to Kant and Rickert, there are two logics (the Aristotelian for analytic propositions, where we can use Venn diagrams and the square of opposition, and the non-Aristotelian for synthetic propositions without Venn diagrams manipulations). This distinction entails another distinction (proposed first by Wilhelm Windelband and Heinrich Rickert) between two kinds of sciences: natural sciences (*Naturwissenschaften*) and cultural sciences (*Geisteswissenschaften*). In the first the Aristotelian logic is used by applying a nomothetic approach, in the second the non-Aristotelian by applying an idiographic approach. The idiographic approach is concerned with individual phenomena, as in biography and much of history, while its opposite, the nomothetic approach, aims to formulate laws as general propositions.

In the history as in an individualizing science (*eine individualisierende Wissenschaft*) we obtain the direct relation between content and extent of historical concepts: the more general historical concept is more value relevant at the same time:

> It obviously remains true that in characterizing one part of historical activity, we can speak of a re-creative understanding of the "spiritual" [der geistigen] world. But the concepts of understanding and re-creation are also too imprecise and general to provide a fully autonomous and exhaustive characterization of the nature of all historical representation. As regards understanding, it is important, first, that the object of understanding in history is always something more than merely real; namely, it is value relevant and meaningful. And second, to remain within the domain of history, the value relevant and the meaningful are comprehended not in a generalizing fashion but in an individualizing fashion, even though their content may be only relatively historical. Finally, even the concept of the "re-creation" of historical individuality acquires its precise significance for the theory of the historical sciences only on the basis of the concept of the individualizing understanding of meaning [7].

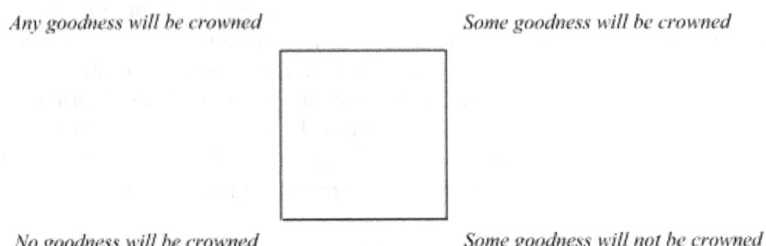

Figure 3. An example of the square of opposition for individualizing propositions

Under these conditions, the general does not imply the particular. For instance, as we saw, we cannot differ the universal proposition 'all bodies are heavy' from the particular one 'some bodies are heavy' by Venn diagrams, because the extents of their concepts (the extents of 'bodies' and 'heavy') are just intersected.

Rickert did not think of creating a new square of opposition that may become suitable for describing semantic opposition between synthetic (historical, individualizing) propositions. If we set up such a problem, we will start in distinguishing between general and particular synthetic propositions.

For analytic propositions while we move from the general (i.e. the concept with the larger extent and the smaller content) to the particular (i.e. the concept with the smaller extent and the larger content), we are losing definiteness and certainty. For synthetic propositions, the general and the particular are two different points of view, because both have different extents and different contents with the same certainty, which satisfy a direct relation between them.

In the Apuleian square of opposition there is a duality between the general and the particular. Indeed, for the general there is a contrary negation and for the particular a subcontrary negation, thereby the contrary negation tends to be maximized and the subcontrary negation tends to be minimized. In the new square of opposition (see Figure 4) we could propose another *duality that takes place between the affirmation and the negation.* In this case the contrary negation holds between the general affirmative proposition and the particular affirmative proposition and the subcontrary negation between the general negative proposition and the particular negative proposition.

Let us try to explain the meaning of the new square. 'Any goodness will be crowned' has no direct meaning, it is a kind of performative proposition expressing immaterial things. Now and here in one and the same context of utterance 'any goodness will be crowned' and 'some goodness will be crowned' cannot be simultaneously true, but can be false. The matter is that both propositions have a different performative meaning (in them I accept goodness differently). Another example is to consider the general proposition 'take my money and buy what you want' and its particular case 'take my money and buy a ring'. Both are contrary, although the first is general and the second is its particular. They express different perlocutions (actions of my hearer after listening), namely the first proposition concerns my care and the second my order, the first proposition allows my hearer to do anything and the second to follow just one action. If after the first proposition the hearer buys the ring, it is not the same as if (s)he buys it after the second proposition.

In new syllogistics we have the following true implication: 'if all bodies are heavy, then no bodies are heavy'. It is senseless for Kant. Due to the transcendental unity of apperception he considers synthetic propositions in the logical way of analytic ones. But it is not correct, synthetic (or performative) propositions have another meaning than analytic propositions. However, in his transcendental dialectics Kant accepts that we have the following true implication with two simple synthetic propositions: 'if the universe is infinite, than it is not infinite'. In this paper synthetic propositions are regarded as performative ones.

In next sections we are formally proving that there exist two squares of opposition and, correspondingly, two syllogistics (the first for analytic propositions and the second for synthetic [performative] propositions).

3 Synthetic syllogictics

In Aristotle's syllogistics, analytic propositions in Kant's words are formalized. In such propositions, a subject is thought within a predicate, the more common concept, for example: 'Socrates is a *man*' or 'All people are *animals*'. Therefore the predicate in formulas SaP ('every S is P'), SiP ('some S are P'), SeP ('no S is P'), SoP ('some S are not P') may be replaced by nouns, but not by adjectives. In other words, we have the following grammar: 'noun 1 (subject) + is + noun 2 (predicate)', and the concept of 'noun 1' is a kind (particular) of the concept of 'noun 2', i.e. 'noun 2' is thought as general for 'noun 1'.

However, how far can we consider propositions like 'Socrates is white', 'All bodies are heavy' within a conventional syllogistics formalizing just analytic propositions? At the first blush, these troubles might be deleted if we transformed an appropriate adjective into a noun. For example, the proposition 'Socrates is white' may be converted to the proposition 'Socrates is a white being', and 'All bodies are heavy' to 'All bodies are something heavy'. However, such a transformation does not solve our problem, because the predicate is not general for the subject still. Thus, Socrates' whiteness is not his substantial attribute and the general of bodies is space, but not weight. Any body is thought first as space entity.

For the first time, Aristotle noticed that there are propositions that have been called synthetic since Kant and these propositions cannot be used in syllogistics. Aristotle's counterexample was as follows: "He who sits, writes and Socrates is sitting, then Socrates is writing" (*Topics* 10). This wrong syllogism was caused by using synthetic propositions. Kant's key example of synthetic propositions is 'All bodies are heavy'. They have the following grammar: 'noun 1 (subject) + is + adjective (empiric or performative attribute)'. The connective '... is ...' of synthetic propositions is understood in this paper as follows: $A \text{ is } B \equiv (\exists C (C \text{ is } A) \land \forall C \forall D ((C \text{ is } A \land D \text{ is } A) \Rightarrow C \text{ is } D) \land \forall C (C \text{ is } A \land C \text{ is } B))$. This understanding corresponds to the Kantian-Rickertian approach.

Assume that all the syllogistic synthetic propositions have the following meaning:

- *Affirmative synthetic a priori* or *general affirmative performance*: 'All S are P' ("All bodies are heavy"): there exist A such that $A \text{ is } S$ and for any A, $A \text{ is } S$ and $A \text{ is } P$;

- *Affirmative synthetic a posteriori* or *particular affirmative performance*: 'Some S are P' ("Socrates is white"): for any A, both $A \text{ is } S$ is false and $A \text{ is } P$ is false;

- *Negative synthetic a priori* or *general negative performance*: 'No S are P' ("No bodies are angels"): there exist A such that $A \text{ is } S$ or $A \text{ is } P$;

- *Negative synthetic a posteriori* or *particular negative performance*: 'Some S are not P' ("Socrates is not black"): for any A, $A \text{ is } S$ is false or there exist A such that $A \text{ is } S$ is false or $A \text{ is } P$ is false.

Let us propose now the syllogistic system formalizing synthetic propositions. This system is said to be *synthetic syllogistics*, while we are assuming that Aristotelian syllogistic is analytic. The basic logical connectives of synthetic syllogistic are as follows: $\mathfrak{a}$ ('every + noun + is + adjective expressing an empiric or performative property'), $\mathfrak{i}$ ('some + noun + is + adjective expressing an empiric or performative property'), $\mathfrak{e}$ ('no + noun

+ is + adjective expressing an empiric or performative property') and o ('some + noun + is not + adjective expressing an empiric or performative property') that are defined in synthetic ontology in the following way:

$$SaP := (\exists A(A \text{ is } S) \wedge (\forall A(A \text{ is } S \wedge A \text{ is } P))); \tag{1}$$

$$SiP := \forall A(\neg(A \text{ is } S) \wedge \neg(A \text{ is } P)); \tag{2}$$

$$SoP := \neg(\exists A(A \text{ is } S) \wedge (\forall A(A \text{ is } S \wedge A \text{ is } P))), \text{ i.e.}$$
$$(\forall A \neg(A \text{ is } S) \vee \exists A(\neg(A \text{ is } S) \vee \neg(A \text{ is } P))); \tag{3}$$

$$SeP := \neg \forall A(\neg(A \text{ is } S) \wedge \neg(A \text{ is } P)), \text{ i.e. } \exists A(A \text{ is } S \vee A \text{ is } P). \tag{4}$$

Now let us formulate axioms of synthetic syllogistics:

$$SaP \Rightarrow SeP; \tag{5}$$

$$SaP \Rightarrow PaS; \tag{6}$$

$$SiP \Rightarrow PiS; \tag{7}$$

$$SaM \Rightarrow SeP; \tag{8}$$

$$MaP \Rightarrow SeP; \tag{9}$$

$$(MaP \wedge SaM) \Rightarrow SaP; \tag{10}$$

$$(MiP \wedge SiM) \Rightarrow SiP. \tag{11}$$

In synthetic syllogistics we have a novel square of opposition that we call the *synthetic square of opposition* (see Figure 3), where the following theorems are inferred: $SaP \Rightarrow \neg(SoP)$, $\neg(SoP) \Rightarrow SaP$, $SiP \Rightarrow \neg(SeP)$, $\neg(SeP) \Rightarrow SiP$, $SeP \Rightarrow \neg(SiP)$, $\neg(SiP) \Rightarrow SeP$, $SoP \Rightarrow \neg(SaP)$, $\neg(SaP) \Rightarrow SoP$, $SaP \Rightarrow \neg(SiP)$, $SiP \Rightarrow \neg(SaP)$, $\neg(SeP) \Rightarrow SoP$, $\neg(SoP) \Rightarrow SeP$, $SaP \Rightarrow SeP$, $SiP \Rightarrow SoP$, $SeP \vee SiP$, $\neg(SeP \wedge SiP)$, $SaP \vee SoP$, $\neg(SaP \wedge SoP)$, $\neg(SaP \wedge SiP)$, $SeP \vee SoP$.

4 Non-Archimedean models of Aristotelian syllogistics and synthetic syllogistics

Suppose B is a complete Boolean algebra with the bottom element 0 and the top element 1 such that the cardinality of its domain $|B|$ is an infinite number. Build up the set B^B of all functions $f: B \mapsto B$. The set of all complements for finite subsets of B is a filter and it is called a Fréchet filter, it is denoted by $\mathcal{U}$. Further, define a new relation $\approx$ on the set B^B by $f \approx g = \{a \in B: f(a) = g(a)\} \in \mathcal{U}$. It is easily proved that the relation

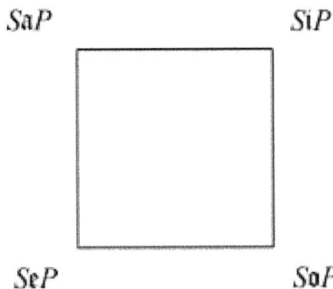

Figure 4. The synthetic square of opposition (for synthetic syllogistics, where synthetic propositions are formalized)

$\approx$ is an equivalence. For each $f \in B^B$ let $[f]$ denote the equivalence class of f under $\approx$. The ultrapower $B^B/\mathcal{U}$ is then defined to be the set of all equivalence classes $[f]$ as f ranges over B^B. This ultrapower is called a *nonstandard* (or *non-Archimedean*) *extension of Boolean algebra* B. It is denoted by $^*B$.

There exist two groups of members of $^*B$: (1) functions that are constant, e.g. $f(a) = m \in B$ on the set $\mathcal{U}$, a constant function $[f = m]$ is denoted by $^*m$, (2) functions that are not constant. The set of all constant functions of $^*B$ is called standard set and it is denoted by $^\circ B$. The members of $^\circ B$ are called standard. It is readily seen that B and $^\circ B$ are isomorphic.

We can extend the usual partial order structure on B to a partial order structure on $^\circ B$:

1. for any members $x, y \in B$ we have $x \leq y$ in B iff $^*x \leq {^*y}$ in $^\circ B$,

2. each member $^*x \in {^\circ B}\backslash\{^*0\}$ (i.e. that is not a bottom element ^{*0} of $^\circ B$) is greater than any number $[f] \in {^*B}\backslash{^\circ B}$ or equal to it, i.e. $^*x \geq [f]$ for any $x \in B$, where $[f]$ is not constant function,

3. ^{*0} is the bottom element of $^*B$.

Notice that under these conditions, there exist the top element $^*1 \in {^*B}$ such that $1 \in B$ and the bottom element $^*0 \in {^*B}$ such that $0 \in B$.

The ordering conditions mentioned above have the following informal sense: (1) the sets $^\circ B$ and B have isomorphic order structure; (2) the set $^*B\backslash\{^*0\}$ contains actual infinities that are less than any member of $^\circ B\backslash\{^*0\}$ (or equal to them). These members are called *Boolean infinitesimals*.

Introduce three operations 'sup', 'inf', '¬' in the partial order structure of $^*B$:

$$\inf([f],[g]) = [\inf(f,g)];$$
$$\sup([f],[g]) = [\sup(f,g)];$$
$$\neg[f] = [\neg f].$$

This means that a nonstandard extension $^*B$ of a Boolean algebra B preserves the least upper bound 'sup', the greatest lower bound 'inf', and the complement '¬' of B.

Consider the member $[h]$ of $^*B$ such that $\{a \in B \colon h(a) \leq f(\neg a)$ or $h(a) > f(\neg a)\} \in \mathcal{U}$. Denote $[h]$ by $[f\neg]$. Then we see that $\inf([f],[f\neg]) \geq {}^*0$ and $\sup([f],[f\neg]) \leq {}^*1$. Really, we have several cases.

1. *Case 1.* The members $\neg[f]$ and $[f\neg]$ are incompatible. Then $\inf([f], [f\neg]) \geq {}^*0$ and $\sup([f],[f\neg]) \leq {}^*1$,

2. *Case 2.* Suppose $\neg[f] \geq [f\neg]$. In this case $\inf([f],[f\neg]) = {}^*0$ and $\sup([f],[f\neg]) \leq {}^*1$.

3. *Case 3.* Suppose $\neg[f] \leq [f\neg]$. In this case $\inf([f],[f\neg]) \geq {}^*0$ and $\sup([f],[f\neg]) = {}^*1$.

4. *Case 4.* The members $[f]$ and $\neg[f\neg]$ are incompatible. Then $\inf(\neg[f], \neg[f\neg]) \geq {}^*0$ and $\sup(\neg[f], \neg[f\neg]) \leq {}^*1$,

5. *Case 5.* Suppose $\neg[f\neg] \geq [f]$. In this case $\inf(\neg[f], \neg[f\neg]) \geq {}^*0$ and $\sup(\neg[f], \neg[f\neg]) = {}^*1$.

6. *Case 6.* Suppose $\neg[f\neg] \leq [f]$. In this case $\inf(\neg[f], \neg[f\neg]) = {}^*0$ and $\sup(\neg[f], \neg[f\neg]) \leq {}^*1$.

7. *Case 7.* The members $\neg[f\neg]$ and $\neg[f]$ are incompatible. Then $\inf([f], \neg[f\neg]) \geq {}^*0$ and $\sup([f], \neg[f\neg]) \leq {}^*1$,

8. *Case 8.* Suppose $\neg[f] \geq \neg[f\neg]$. In this case $\inf([f], \neg[f\neg]) = {}^*0$ and $\sup([f], \neg[f\neg]) \leq {}^*1$.

9. *Case 9.* Suppose $\neg[f] \leq \neg[f\neg]$. In this case $\inf([f], \neg[f\neg]) \geq {}^*0$ and $\sup([f], \neg[f\neg]) = {}^*1$.

10. *Case 10.* The members $[f]$ and $[f\neg]$ are incompatible. Then $\inf(\neg[f], [f\neg]) \geq {}^*0$ and $\sup(\neg[f], [f\neg]) \leq {}^*1$,

11. *Case 11.* Suppose $[f\neg] \geq [f]$. In this case $\inf(\neg[f], [f\neg]) \geq {}^*0$ and $\sup(\neg[f], [f\neg]) = {}^*1$.

12. *Case 12.* Suppose $[f\neg] \leq [f]$. In this case $\inf(\neg[f], [f\neg]) = {}^*0$ and $\sup(\neg[f], [f\neg]) \leq {}^*1$.

Definition 1. Now define hyperrational valued matrix logic $\mathfrak{M}_B$ as the ordered system $\langle {}^*B, \{{}^*1\}, \neg, \Rightarrow, \vee, \wedge \rangle$, where

1. ${}^*B$ is the set of truth values,
2. $\{{}^*1\}$ is the set of designated truth values,
3. for all $[x] \in {}^*B$, $\neg[x] = {}^*1 - [x]$,
4. for all $[x], [y] \in {}^*B$, $[x] \Rightarrow [y] = {}^*1 - \sup([x], [y]) + [y]$,
5. for all $[x], [y] \in {}^*B$, $[x] \wedge [y] = \inf([x], [y])$,
6. for all $[x], [y] \in {}^*B$, $[x] \vee [y] = \sup([x], [y])$.

Proposition 1. In $\mathfrak{M}_B$ there are only two squares of opposition.

Proof. We have just eight cases: (1) $[f] \leq [f\neg]$, (2) $[f] \leq \neg[f\neg]$, (3) $[f\neg] \leq [f]$, (4) $[f\neg] \leq \neg[f]$, (5) $\neg[f\neg] \leq [f]$, (6) $\neg[f\neg] \leq \neg[f]$, (7) $\neg[f] \leq [f\neg]$, (8) $\neg[f] \leq \neg[f\neg]$. Taking into account that couples $[f]$ and $\neg[f]$ ($[f\neg]$ and $\neg[f\neg]$) are contradictory, we can claim that there exist two squares of opposition:

- if $[f] \leq \neg[f\neg]$ (resp. $[f\neg] \leq \neg[f]$), we have the conventional square of opposition (see Figure 5); if $\neg[f\neg] \leq [f]$ (resp. $\neg[f] \leq [f\neg]$), we have its dual without changing meaning. □

- if $[f\neg] \leq [f]$ (resp. $\neg[f] \leq \neg[f\neg]$), we have the synthetic square of opposition; if $[f] \leq [f\neg]$ (resp. $\neg[f\neg] \leq \neg[f]$), we have its dual without changing meaning (see Figure 6);

Now we can build models for atomic syllogistic formulas (i.e. syllogistic formulas without propositional connectives) due to algebra $\mathfrak{M}_B$.

Definition 2. A structure $\mathfrak{B} = \langle O, I, \mathsf{à}, \mathsf{è}, \mathsf{ì}, \mathsf{ò}, \mathsf{ȧ}, \mathsf{ė}, \mathsf{i}, \mathsf{ȯ} \rangle$ is a *non-Archimedean syllogistic model* iff:

1. O is a restriction of the set $\mathfrak{M}_B$ to an appropriate square (triangle) of opposition (thereby the conventional square of opposition should hold true for Aristotelian syllogistics and the synthetic square of opposition holds for synthetic syllogistics).

2. I is a mapping that associates a class of equivalence $[f] \in O$ with each atomic syllogistic formula $S \diamond P$, where $\diamond \in \{\mathsf{a}, \mathsf{e}, \mathsf{i}, \mathsf{o}, \mathsf{a}, \mathsf{e}, \mathsf{i}, \mathsf{o}\}$, so that $I(S \diamond P) = |S|\diamond|P|$, where $\dot\diamond \in \{\mathsf{à}, \mathsf{è}, \mathsf{ì}, \mathsf{ò}, \mathsf{ȧ}, \mathsf{ė}, \mathsf{i}, \mathsf{ȯ}\}$ and

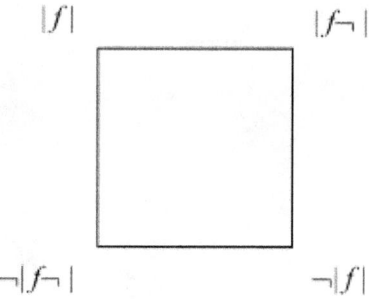

Figure 5. In case $[f\neg] \leq \neg[f]$, the square of opposition for any members $[f]$, $[f\neg]$, $\neg[f]$, $\neg[f\neg]$ of $^*B$ holds true, i.e. $[f], [f\neg]$ are contrary, $[f], \neg[f]$ (resp. $\neg[f\neg], [f\neg]$) are contradictory, $\neg[f\neg], \neg[f]$ are subcontrary, $[f], \neg[f\neg]$ (resp. $[f\neg], \neg[f]$) are said to stand in the subalternation.

- $|S|\dot{\mathbf{a}}|P| = [f]$ (resp. $|S|\dot{\mathbf{a}}|P| = \neg[f\neg]$);
- $|S|\dot{\mathbf{e}}|P| = [f\neg]$ (resp. $|S|\dot{\mathbf{e}}|P| = \neg[f]$);
- $|S|\dot{\mathbf{i}}|P| = \neg[f\neg]$ (resp. $|S|\dot{\mathbf{i}}|P| = [f]$);
- $|S|\dot{\mathbf{o}}|P| = \neg[f]$ (resp. $|S|\dot{\mathbf{o}}|P| = [f\neg]$);
- $|S|\ddot{\mathbf{a}}|P| = [f]$ (resp. $|S|\ddot{\mathbf{a}}|P| = [f\neg]$);
- $|S|\ddot{\mathbf{e}}|P| = [f\neg]$ (resp. $|S|\ddot{\mathbf{e}}|P| = [f]$);
- $|S|\ddot{\mathbf{i}}|P| = \neg[f\neg]$ (resp. $|S|\ddot{\mathbf{i}}|P| = \neg[f]$);
- $|S|\ddot{\mathbf{o}}|P| = \neg[f]$ (resp. $|S|\ddot{\mathbf{o}}|P| = \neg[f\neg]$).

We now give the truth conditions of Boolean combinations of atomic syllogistic formulas in a non-Archimedean syllogistic model:

Definition 3.

$\mathfrak{B} \models \neg \phi$ iff $\mathfrak{B} \not\models \phi$
$\mathfrak{B} \models \phi \wedge \psi$ iff $\mathfrak{B} \models \phi$ and $\mathfrak{B} \models \psi$
$\mathfrak{B} \models \phi \vee \psi$ iff $\mathfrak{B} \models \phi$ or $\mathfrak{B} \models \psi$
$\mathfrak{B} \models \phi \Rightarrow \psi$ iff $\mathfrak{B} \models \neg\phi$ or $\mathfrak{B} \models \psi$

5 Performative propositions as a kind of synthetic propositions

Performative propositions are said to be propositions that contain performative verbs, i.e. verbs that express unreal, imaginary actions such as 'think',

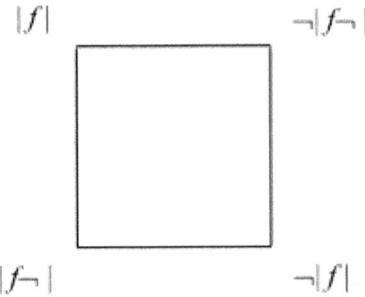

Figure 6. In case $\neg[f\neg] \leq \neg[f]$, the *synthetic* square of opposition for any members $[f]$, $[f\neg]$, $\neg[f]$, $\neg[f\neg]$ of $^*B$ holds true, i.e. $[f], \neg[f\neg]$ are contrary, $[f], \neg[f]$ (resp. $\neg[f\neg], [f\neg]$) are contradictory, $\neg[f], [f\neg]$ are subcontrary, $[f], [f\neg]$ (resp. $\neg[f\neg], \neg[f]$) are said to stand in the subalternation.

'feel', 'hope', 'ask', etc.

Furthermore, if we keep the following hypothesis, we could show that in fact performative propositions satisfy the synthetic square of opposition.

Hypothesis 1. Suppose $\mathbf{F}_1(\Phi) \supset \mathbf{F}_2(\Phi)$ is successful and $\mathbf{F}'_1(\Phi) \supset \mathbf{F}'_2(\Phi)$ is successful. The same speaker cannot successfully utter two different performative propositions $\mathbf{F}_1(\Phi)$ and $\mathbf{F}'_2(\Phi)$ at the same time and at the same place. One of the two propositions $\mathbf{F}_1(\Phi)$, $\mathbf{F}'_2(\Phi)$ or both are epicly failed.

Let us consider an example. Assume the four performative propositions are ordered as follows:

('I *forbid* not to do') $\supset$ ('I *permit* to do'),

('I *order* to do') $\supset$ ('I *beg* to do'),

('I *am sure* not to do') $\supset$ ('I *make doubt* about doing'), etc.

Then we find out the following four distinct kinds of opposition between pairs of performative propositions:

1. First, performative verbs '*forbid not to do*' and '*make doubt about doing*' (resp. '*order to do*' and '*permit to do*', '*be sure to do*' and '*beg to do*', etc.) are *contrary*. They cannot be successful together at the same time, at the same place, and with the same propositional content. However, both may be unsuccessful.

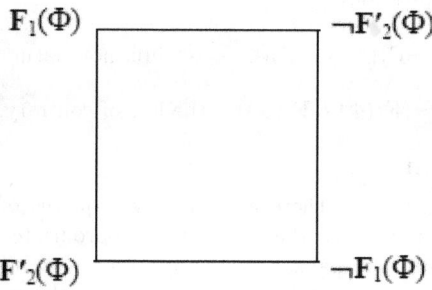

Figure 7. The synthetic square of opposition for performative verbs $\mathbf{F}_1$, $\mathbf{F}'_2$. For instance, $\mathbf{F}_1$ is 'order to do' and $\mathbf{F}'_2$ is 'permit not to do.' Then $\neg \mathbf{F}_1$ is 'ask not to do' and $\neg \mathbf{F}'_2$ is 'forbid not to do.'

2. Second, performative verbs *'forbid not to do'* and *'not forbid not to do,'* i.e. *'permit to do,'* (resp. *'make doubt about doing'* and *'not make doubt about doing'*, i.e. *'be sure to do,' 'order to do'* and *'not order to do,'* i.e. *'ask not to do,'* etc.) are *contradictory*, i.e. the success of the one implies the unsuccess of the other, and conversely.

3. Further, performative verbs *'ask not to do'* and *'forbid not to do'* (resp. *'be sure not to do'* and *'permit not to do,'* etc.) are *subcontrary*, i.e. it is impossible for both to be unsuccessful in corresponding performative propositions at the same time, at same place, and with the same propositional content, however it is possible both to be successful.

4. Finally, performative verbs *'order to do'* and *'forbid not to do'* (resp. *'permit not to do'* and *'ask not to do'*) are said to stand in the *subalternation*, i.e. the success of the first (the superaltern) implies the success of the second (the subaltern), but not conversely.

Hence, if we have two ordered pairs of atomic performative propositions such as $\mathbf{F}_1(\Phi) \supset \mathbf{F}_2(\Phi)$ and $\mathbf{F}'_1(\Phi) \supset \mathbf{F}'_2(\Phi)$, then we could construct the synthetic square of opposition over them, see Figure 7. Indeed, then the pair $\mathbf{F}_1(\Phi)$ and $\mathbf{F}'_2(\Phi)$ is called contraries (contrariae), the pairs $\mathbf{F}_1(\Phi)$ and $\neg\mathbf{F}_1(\Phi)$, $\mathbf{F}'_2(\Phi)$ and $\neg\mathbf{F}'_2(\Phi)$ contradictories (contradictoriae), the pair $\neg\mathbf{F}_1(\Phi)$ and $\neg\mathbf{F}'_2(\Phi)$ subcontraries (subcontrariae), and the pairs $\mathbf{F}_1(\Phi)$ and $\neg\mathbf{F}'_2(\Phi)$, $\mathbf{F}'_2(\Phi)$ and $\neg\mathbf{F}_1(\Phi)$ subalternates (subalternae).

In the synthetic square of opposition, the following expressions are true performative propositions:

$$\neg \mathbf{F}_1(\Phi) \vee \neg \mathbf{F}_2(\Phi) - \text{tertium non datur}, \qquad (12)$$

$$\neg(\mathbf{F}_1(\Phi) \wedge \mathbf{F}_2(\Phi)) - \text{the law of contrary}. \qquad (13)$$

6 Conclusion

Proposition 1 states that there are only two squares of opposition if we assume Boolean algebra as the basis of an appropriate non-Archimedean extension. The conventional square of opposition may be aimed for getting analytic syllogistics (Aristotelian syllogistics) and the new one for getting synthetic syllogistics (syllogistics, proposed in this paper). Also, we can define a novel duality relation between conjunction and disjunction, universal quantifier and existential quantifier, modal operators of necessity and possibility and, as a result, we can build out absolutely new first-order logic and modal logic.

BIBLIOGRAPHY

[1] Apuleius. *Pro Se De Magia (Apologia)*. Vincent Hunink (editor). Amsterdam: Gieben, 1977.
[2] Apuleius. *The Logic of Apuleius: including a complete Latin text and English translation of the Peri hermeneias of Apuleius of Madaura*. David Londey and Carmen Johanson (ed. and trans.). Leiden, New York: E.J. Brill, 1987.
[3] Bocheński, Innocenty M. *Formale Logik*, Freiburg-München: Karl Alber, 1956.
[4] Kant, I., Critique of Pure Reason. Cambridge University Press, 1999.
[5] Robinson, A. *Non-Standard Analysis. Studies in Logic and the Foundations of Mathematics*. North-Holland, 1966.
[6] L. E. Rose. *Aristotle's Syllogistic*. Charles C. Thomas Publisher, 1968.
[7] H. Rickert. *Die Grenzen der Naturwissenschaftlichen Begriffsbildung. Eine logische Einleitung in die historischen Wissenschaften*. 2. Aufl. Mohr: Tubingen, 1913.
[8] W. D. Ross (editor), *The Works of Aristotle*, Volume 1: Logic. Oxford University Press, 1928.
[9] Sullivan, Mark. *Apuleian logic*. Amsterdam, North-Holland Pub. Co., 1967.

Andrew SCHUMANN
University of Information Technology and Management in Rzeszow,
Rzeszow, Poland

What Is a Refutation System?

Tomasz Skura

1 Introduction

A refutation system is an axiomatic system for non-valid formulas. It consists of refutation axioms, which are some non-valid formulas, and refutation rules, which are some rules preserving non-validity. (The formal concept was introduced by Łukasiewicz [1], but the idea was already known to Aristotle.) A formula A is *refutable* iff there is a derivation of A from refutation axioms by refutation rules. Refutations are proofs.

Why should anyone be interested in refutation systems?

First, axiomatic systems are easy to understand. All you have to know is what a derivation is. (It is a finite sequence of formulas in which every formula is an axiom or is obtained from preceding formulas by a rule.)

Second, our syntactic completeness proofs are exact, constructive, and elegant.

Third, refutation systems may provide interesting syntactic characterizations of non-classical logics resulting from Classical Logic by rejecting some classical laws, for example paraconsistent logics (see [3,4]).

Fourth, refutation systems (with certain normal forms) provide a method of axiomatizing both a logic L and its complement $-L$ (see [5]).

Fifth, refutation systems may have practical applications. They may provide simple decision procedures.

It is natural to ask how refutation systems relate to other standard refutation methods (models and tableaux). In this paper we deal with this question. We choose, by way of illustration, Classical Propositional Logic (CL), which we define as the set of all Boolean laws.

It turns out that syntactic refutations (of a certain kind) and countermodels are two sides of one coin, whereas refutation systems and tableau systems are complementary. Also, refutation systems are capable not only of refutation search but also of proof search. What is more, in specific situations, refutation procedures may be simpler than tableau procedures.

2 Refutations and Models

By a *model* we mean a Boolean valuation v assigning either 1 or 0 to every propositional variable and extended to the other formulas in the standard way, that is, $v(A \wedge B) = 1$ iff $v(A) = 1$ and $v(B) = 1$, and so on. We say that v is a counter-model for A iff $v(A) = 0$.

Every refutation determines a counter-model, and every counter-model determines a refutation. Let us make these ideas more precise.

We assume that we have a refutation system of the following kind.

Refutation axioms: Some formulas $F \notin CL$.
Refutation rules: Some rules of the type
B/A, where $A \to B \in CL$.
Note that if v is a valuation and $v(B) = 0$, then $v(A) = 0$.

Let F be a refutation axiom. By an F-refutation of a formula A we mean a finite sequence $A_1, ..., A_n$ such that.
(i) $A_1 = A$.
(ii) $A_n = F$.
(iii) Every A_i with $i < n$ is obtained from a successor by a refutation rule.

We also define v_F to be a valuation v such that $v(F) = 0$.

Theorem 1 *Let F be a refutation axiom. If A has an F-refutation, then $v_F(A) = 0$.*
PROOF Assume that A has an F refutation $A_1, ..., A_n$. By definition
$$v_F(A_n) = 0.$$
Hence $v_F(A_{n-1}) = 0$, so ..., so $v_F(A_1) = 0$, which means that $v_F(A) = 0$, as required.

In order to show that every counter-model determines a refutation, for any model v we define the corresponding refutation system $\mathbf{R}_v$.

As usual, for any finite sets X, Y of formulas the symbol $X \longrightarrow Y$ stands for $\bigwedge X \to \bigvee Y$, where $\bigwedge X$ ($\bigvee Y$) is a conjunction (a disjunction) of the formulas in X (Y). Also, $\bigwedge \emptyset = \top$ and $\bigvee \emptyset = \bot$. (Of course, $v(\top) = 1$ and $v(\bot) = 0$ for every valuation v.)

We shall be using the symbol $X \longrightarrow$ instead of $X \longrightarrow \emptyset$ (which stands for $\bigwedge X \to \bot$) and $\longrightarrow$ for $\emptyset \longrightarrow$ (which is $\top \to \bot$). We also write $X, Y \longrightarrow Z$ for $X \cup Y \longrightarrow Z$ and $X, A_1, ..., A_n \longrightarrow$ for $X, \{A_1, ..., A_n\} \longrightarrow$.

Recall that for every formula A there is a conjunctive normal form B such that $A \equiv B \in CL$ and B is a conjunction $B_1 \wedge ... \wedge B_m$, where each $B_i = X \longrightarrow Y$ and X, Y are finite sets of propositional variables.

Our refutation system $\mathbf{R}_v$ is the following.

Refutation axioms: $X \longrightarrow Y$, where X, Y are finite sets of propositional variables such that $v(X \longrightarrow Y) = 0$.
Refutation rules:

$(R_{CNF}) \quad \dfrac{CNF(A)}{A}$ where $CNF(A)$ is a conjunctive normal form for A

$(R_\wedge) \quad \dfrac{B_i}{B_1 \wedge ... \wedge B_m} \quad (1 \leq i \leq m)$

Theorem 2 *If v is a counter-model for A, then A is refutable in $\mathbf{R}_v$.*
PROOF Assume that $v(A) = 0$. Let $CNF(A) = B_1 \wedge ... \wedge B_m$. We have $v(B_i) = 0$ for some i. Also, $B_i = X \longrightarrow Y$ for some finite sets X, Y of variables, so B_i is a refutation axiom. Hence $CNF(A)$ is refutable by $R_\wedge$, and so A is refutable by R_{CNF}.

3 Refutations and Tableaux

In a tableau procedure, you assume that A is non-valid, so $\neg A$ is satisfiable. Then you apply (branching) tableau rules preserving satisfiability. If every branch closes (is inconsistent), then a proof for A is found. This is a proof-search procedure.

And in a refutation procedure, you assume that A is valid and you apply inverse refutation rules preserving validity. If you arrive at a refutation axiom (which is non-valid), then a refutation for A is found. This is a refutation-search procedure.

Thus, refutation systems and tableau systems are complementary.

Of course, it is a standard approach to use tableau systems not only for proofs but also for refutations. If no tableau for $\neg A$ closes, then a counter-model for A can be constructed. We have the following decision procedure.

(Tableaux) Either you get a proof for A or you get a counter-model for A.

Refutation systems are also capable of proof search, if they are of the following kind. The refutation rules are invertible and they have the property that each premise is simpler than the conclusion. Then we have the following decision procedure.

(Refutations) Either you get a refutation for A or you get a proof for A.

Thus, both methods provide decision procedures. However, tableau procedures involve **repetitions**, while our refutation procedures consist in **reductions**, so that in a specific situation our procedure may be simpler. We illustrate this by introducing a useful refutation system for CL providing a neat decision procedure.

4 A Useful Refutation System

We start with some definitions. Let a be a propositional variable. We put $a^* = \neg a$ and $(\neg a)^* = a$. Both a and $\neg a$ are called *literals*. A *clause* is a disjunction $\bigvee X$, where X is a finite non-empty set of literals. We also write $A_1 \vee ... \vee A_n$ for $\bigvee\{A_1, ..., A_n\}$. Note that the order in a clause $A_1 \vee ... \vee A_n$ is ignored and repetitions are not allowed (so $p \vee p$ is not a clause but p is).

By a *normal form* we mean a formula
$$\Gamma \longrightarrow$$
where Γ is a finite set of clauses. Every formula A can be reduced to a normal form A' with the following property.
$$A \in CL \text{ iff } A' \in CL.$$
(For more information see e.g. [2].)

The *rank* of a normal form $\Gamma \longrightarrow$ is the number of literals l such that l and l^* occur in distinct clauses in Γ. For example, the rank of
$$p \vee \neg p, \neg q, q \vee r, t \vee \neg s, t \longrightarrow$$
is 1.

We now introduce the following refutation system for CL.

Refutation axioms: All normal forms of rank 0.
Refutation rules:

(R_1) $\qquad \dfrac{l, A_1, ..., A_n, \Gamma \longrightarrow}{l, l^* \vee A_1, ..., l^* \vee A_n, \Gamma \longrightarrow}$

(R_2) $\qquad \dfrac{\{A_i \vee B_j : 1 \leq i \leq m, 1 \leq j \leq n\}, \Delta \longrightarrow}{l \vee A_1, ..., l \vee A_m, l^* \vee B_1, ..., l^* \vee B_n, \Delta \longrightarrow}$

provided that l, l^* do not occur in Δ.

It is easy to see that the refutation axioms are not in CL (just assign 1 to l for every $l \vee \bigvee X \in \Gamma$) and the refutation rules preserve non-validity. For example, to check R_2, assume that the premise is not in CL, so there is a Boolean valuation v such that $v(A_i \vee B_j) = 1$ for all i, j and $v(\bigwedge \Delta) = 1$. Either $v(A_i) = 1$ for all i or $v(A_i) = 0$ for some i. If $v(A_i) = 1$ for all i, then $v(l \vee A_i) = 1$ for all i, so by assigning 1 to l^* we define a new valuation

v' such that $v'(l^* \vee B_j) = 1$ for all j and $v'(\bigwedge \Delta) = 1$, and so the conclusion is not in CL. And if $v(A_i) = 0$ for some i, then $v(B_j) = 1$ for all j, so $v(l^* \vee B_j) = 1$ for all j, and so by assigning 1 to l we have a new valuation refuting the conclusion.

Hence if F is refutable, then $F \notin CL$.

Note that the rules are invertible (that is, if the conclusion in non-valid, then so is the premise). For example, assume that the conclusion of R_2 is not in CL. Then there is a valuation v such that $v(l \vee A_i) = v(l^* \vee B_j) = 1$ for all i, j and $v(\bigwedge \Delta) = 1$. Either $v(l) = 0$ or $v(l) = 1$. If $v(l) = 0$, then $v(A_i) = 1$ for all i, so $v(A_i \vee B_j) = 1$. And if $v(l) = 1$, then $v(l^*) = 0$, so $v(B_j) = 1$ for all j, and so $v(A_i \vee B_j) = 1$. Hence the premise is not in CL.

Remark 3 The rule R_2 is not of the kind defined in Section 2. Let $A = p \vee q, \neg p \vee r \longrightarrow$ and $B = q \vee r \longrightarrow$. Then $A \to B \notin CL$. Indeed, let v be a valuation such that $v(r) = 1$, $v(p) = v(q) = 0$. Then $v(A) = 1$ and $v(B) = 0$.

Theorem 4 *For every normal form $F = \Gamma \longrightarrow$, either $F \in CL$ or F is refutable.*

PROOF By induction on the rank $r(F)$ of F.

(i) $r(F) = 0$. Then F is a refutation axiom, so F is refutable.

(ii) $r(F) > 0$ and this is true for normal forms of rank $< r(F)$. Then there is a literal l such that $l \vee \bigvee X \in \Gamma$ and $l^* \vee \bigvee Y \in \Gamma$ and these are distinct clauses.

(Case 1) $l \in \Gamma$.
(Case 1.1) $l^* \in \Gamma$. Then $F \in CL$.
(Case 1.2) $l^* \notin \Gamma$. Then the clauses with l^* have the form $l^* \vee B$. Let
$l^* \vee B_1, ..., l^* \vee B_n$
be all such clauses. Consider the normal form
$F' = \Gamma', B_1, ..., B_n \longrightarrow$
where $\Gamma' = \Gamma - \{l^* \vee B_1, ..., l^* \vee B_n\}$. Note that F' is of rank $< r(F)$, so by the induction hypothesis, either $F' \in CL$ or F' is refutable. If F' is refutable, then so is F by R_1. And if $F' \in CL$, then $F \in CL$ (because R_1 is invertible).

(Case 2) $l \notin \Gamma$. Then the clauses with l have the form $l \vee A$. Let
$l \vee A_1, ..., l \vee A_m$
be all such clauses.

(Case 2.1) $l^* \in \Gamma$. Then proceed as in Case 1.2.

(Case 2.2) $l^* \notin \Gamma$. Then the clauses with l^* have the form $l^* \vee B$. Let $l^* \vee B_1, ..., l^* \vee B_n$ be all such clauses. Consider the normal form
$$F'' = \{A_i \vee B_j : 1 \leq i \leq m, 1 \leq j \leq n\}, \Gamma'' \longrightarrow$$
where $\Gamma'' = \Gamma - \{l \vee A_1, ..., l \vee A_m, l^* \vee B_1, ..., l^* \vee B_n\}$. Since F'' is of rank $< r(F)$, by the induction hypothesis, either $F'' \in CL$ or F'' is refutable. If $F'' \in CL$ then $F \in CL$ (because R_2 is invertible). And if F'' is refutable, then so is F by R_2.

Therefore either $F \in CL$ or F is refutable, which was to be shown.

Hence if $F \notin CL$, then F is refutable, so that we have: $F \notin CL$ iff F is refutable.

Remark 5 The following special case of R_2 seems interesting.

$$(R_2') \quad \frac{A, \Delta \longrightarrow}{l \vee A, l^* \vee A, \Delta \longrightarrow}$$

(Note that Δ can be arbitrary in R_2'.)

Example 6 Let $F = \Gamma \longrightarrow$, where

$$\Gamma = \{p_1 \vee p_2, \neg p_1 \vee p_2,$$
$$\neg p_2 \vee p_3, \neg p_3 \vee p_4, ..., \neg p_{99} \vee p_{100}\}$$

We reduce F to a refutation axiom in 100 steps by using R_2' and R_1.

F
$F_1 = p_2, \neg p_2 \vee p_3, \neg p_3 \vee p_4, ... \longrightarrow$
$F_2 = p_2, p_3, \neg p_3 \vee p_4, ... \longrightarrow$
.
.
.
$F_{99} = p_2, p_3, p_4, ..., p_{99}, p_{100} \longrightarrow$

5 A Neat Decision Procedure

The refutation system in Section 4 provides a decision procedure for CL, but this procedure involves a lot of redundant things. We now simplify the procedure by modifying the refutation system as follows.

Refutation axiom: $\longrightarrow$
Refutation rules:

(S_0) $\quad \dfrac{\Delta \longrightarrow}{C_1, ..., C_k, \Delta \longrightarrow}$

where each clause $C_i = l \vee \bigvee X_i$, and l, l^* do not occur in Δ.

(S_1) $\quad \dfrac{B_1, ..., B_n, \Delta \longrightarrow}{l, \Theta, \Delta \longrightarrow}$

where
$$\Theta = \{l \vee A_1, ..., l \vee A_m, l^* \vee B_1, ..., l^* \vee B_n\}$$
and l, l^* do not occur in Δ.

(S_2) $\quad \dfrac{F_1}{F} \quad \dfrac{F_2}{F}$

where
$F = \Theta, \Delta \longrightarrow$
$F_1 = A_1, ..., A_m, \Delta \longrightarrow$
$F_2 = B_1, ..., B_n, \Delta \longrightarrow$
Θ is as above and l, l^* do not occur in Δ.

We say $\dashv F$ (F is refutable) iff F is derivable in this system.

Note that S_0, S_1 are invertible and we have:
$F \notin CL$ iff $F_1 \notin CL$ or $F_2 \notin CL$ (see the argument justifying R_2).

The completeness proof is a modification of that in Section 4.
By the size $s(F)$ of F we mean the number of propositional variables occurring in F.
We introduce the following proof system for CL.

Proof axioms: $l, l^*, \Gamma \longrightarrow$
Proof rules: S_0, S_1, and

(S_2') $\quad \dfrac{F_1 \quad F_2}{F}$

We say $\vdash F$ (F is provable) iff F is derivable in this system.

Theorem 7 If F is a normal form, then either $\vdash F$ or $\dashv F$.
PROOF By induction on $s(F)$.

(i) $s(F) = 0$. Then $\Gamma = \emptyset$, so F is the refutation axiom, and so $\dashv F$.

(ii) $s(F) > 0$ and this is true for normal forms of size $< s(F)$. Then there is a literal l such that $l \vee \bigvee X \in \Gamma$ for some X.

Let $C_1, ..., C_k$ be all clauses of the form $l \vee \bigvee X$ in Γ, and let $\Gamma_0 = \Gamma - \{C_1, ..., C_k\}$.
(Case 1) l^* does not occur in Γ_0. Consider the normal form
$$F_0 = \Gamma_0 \longrightarrow$$
Since l, l^* do not occur in Γ_0, we have $s(F_0) < s(F)$, so by the induction hypothesis, either $\vdash F_0$ or $\dashv F_0$. If $\vdash F_0$ then $\vdash F$ (by S_0), and if $\dashv F_0$ then $\dashv F$ (by S_0). Therefore either $\vdash F$ or $\dashv F$.

(Case 2) l^* occurs in Γ_0.
(Case 2.1) $l \in \Gamma$.
(Case 2.1.1) $l^* \in \Gamma$. Then F is a proof axiom, so $\vdash F$.
(Case 2.1.2) $l^* \notin \Gamma$. Then the clauses with l^* have the form $l^* \vee B$. Let
$$l^* \vee B_1, ..., l^* \vee B_n$$
be all such clauses. Consider the normal form
$$F' = B_1, ... B_n, \Gamma' \longrightarrow$$
where $\Gamma' = \Gamma_0 - \{l^* \vee B_1, ..., l^* \vee B_n\}$. Note that F' is of size $< s(F)$, so by the induction hypothesis, $\vdash F'$ or $\dashv F'$. If $\vdash F'$ then $\vdash F$ (by S_1), and if $\dashv F'$ then $\dashv F$ (by S_1).

(Case 2.2) $l \notin \Gamma$. Then the clauses with l have the form $l \vee A$. Let
$$l \vee A_1, ..., l \vee A_m$$
be all such clauses.
(Case 2.2.1) $l^* \in \Gamma$. Consider the normal form
$$F' = A_1, ..., A_m, \Gamma' \longrightarrow$$
where Γ' results from Γ by deleting all clauses with l and all clauses with l^*. F' is of size $< s(F)$, so by the induction hypothesis, $\vdash F'$ or $\dashv F'$. If $\vdash F'$ then $\vdash F$ (by S_1), and if $\dashv F'$ then $\dashv F$ (by S_1).

(Case 2.2.2) $l^* \notin \Gamma$. Then the clauses with l^* have the form $l^* \vee B$. Let
$$l^* \vee B_1, ... l^* \vee B_n$$
be all such clauses. Consider the normal forms
$$F_1 = A_1, ..., A_m, \Gamma'' \longrightarrow$$
$$F_2 = B_1, ..., B_n, \Gamma'' \longrightarrow$$
where $\Gamma'' = \Gamma - \{l \vee A_1, ..., l \vee A_m, l^* \vee B_1, ..., l^* \vee B_n\}$. Since F_1, F_2 are of size $< s(F)$, by the induction hypothesis, we have:
$\vdash F_1$ or $\dashv F_1$
$\vdash F_2$ or $\dashv F_2$
If both $\vdash F_1$ and $\vdash F_2$, then $\vdash F$ (by S_2'). And if $\dashv F_1$ or $\dashv F_2$, then $\dashv F$ (by S_2).

Therefore either $\vdash F$ or $\dashv F$, which was to be shown.

We now describe our decision procedure for $F = \Gamma \longrightarrow$.
First, we define the following inverse rules.

$(Inv(S_0))$ $\quad \dfrac{C_1,...,C_k, \Delta \longrightarrow}{\Delta \longrightarrow}$

$(Inv(S_1))$ $\quad \dfrac{l, \Theta, \Delta \longrightarrow}{B_1,..,B_n, \Delta \longrightarrow}$

$(Inv(S_2))$ $\quad \dfrac{F}{F_1 | F_2}$

Second, we define a finite tree for F with finite sets of clauses as nodes satisfying the following conditions.

(1) The origin is Γ.
(2.1) If Y is the immediate successor of a node X, then $Y \longrightarrow$ is obtained from $X \longrightarrow$ by $Inv(S_0)$ or $Inv(S_1)$.
(2.2) If Y, Z are the immediate successors of a node X, then
$Y \longrightarrow \quad | \quad Z \longrightarrow$
are obtained from $X \longrightarrow$ by $Inv(S_2)$.
(3) X is an end node iff either $X \longrightarrow$ is a proof axiom or $X \longrightarrow$ is the refutation axiom.

Note that applying $Inv(S_0), Inv(S_1)$ to X amounts to deleting things in X, and applying $Inv(S_2)$ to X amounts to deleting either all $l \vee A_i$ (and l^*) or all $l^* \vee B_j$ (and l) in X.

In each step one propositional variable is eliminated, so that the tree is indeed finite. Also we have:

$\dashv F$ iff some end node is $\emptyset$, and $\vdash F$ iff no end node is $\emptyset$.

Example 8 Let F be the normal form defined in Example 6. We reduce F to $\emptyset$ as follows.
Γ
$p_2, \neg p_2 \vee p_3, ..., \neg p_{99} \vee p_{100}$ $\quad$ (by $Inv(S_2)$)
$p_3, ..., \neg p_{99} \vee p_{100}$ $\quad$ (by $Inv(S_1)$)
.
.
.
p_{100} $\quad$ (by $Inv(S_1)$)
$\emptyset$ $\quad$ (by $Inv(S_0)$)
Hence $\dashv F$.

Example 9 Let $F = \Gamma \longrightarrow$, where

$\Gamma = \{p_1 \vee p_2, p_1 \vee p_3, ..., p_1 \vee p_{100},$
$\quad p_1 \vee \neg p_{100},$
$\quad \neg p_1 \vee \neg p_2, \neg p_1 \vee \neg p_3, ..., \neg p_1 \vee \neg p_{100},$
$\quad \neg p_1 \vee p_{100}\}$

We get Γ_1, Γ_2 (by $Inv(S_2)$), where
$\Gamma_1 = \{p_2, p_3, ..., p_{100}, \neg p_{100}\}$
$\Gamma_2 = \{\neg p_2, \neg p_3, ..., \neg p_{100}, p_{100}\}$
and they are end nodes. Since no end node is $\emptyset$, we have $\vdash F$.

The tableau procedure for F is not so neat.

References

[1] J. Łukasiewicz, *Aristotle's Syllogistic from the Standpoint of Modern Formal Logic*. Oxford, 1951.

[2] G. Mints, Gentzen-type systems and resolution rules. *Lecture Notes in Computer Science* 417 (1990), 198-231.

[3] T. Skura, The RM paraconsistent refutation system. *Logic and Logical Philosophy* 18 (2009), 65-70.

[4] T. Skura, Refutation systems in propositional logic. In: D. Gabbay and F. Guenthner (eds.) *Handbook of Philosophical Logic*, vol. 16 (2011), 115-157.

[5] T. Skura, On refutation rules. *Logica Universalis* 5 (2011), 249-254.

Tomasz SKURA
Institute of Philosophy
University of Zielona Gora, Poland

www.ingramcontent.com/pod-product-compliance
Lightning Source LLC
Chambersburg PA
CBHW070742160426
43192CB00009B/1541

*9 7 8 1 8 4 8 9 0 0 9 0 5 *